Joseph Priestley, James Basire, William Hey

Observations on Different Kinds of Air

Joseph Priestley, James Basire, William Hey

Observations on Different Kinds of Air

ISBN/EAN: 9783337812072

Printed in Europe, USA, Canada, Australia, Japan

Cover: Foto ©berggeist007 / pixelio.de

XIX *Obſervations on different Kinds of Air. By* Joſeph Prieſtley, *LL. D. F. R. S.*

Read March 5, 12, 19, 26, 1772.

THE following obſervations on the properties of ſeveral different kinds of air, I am ſenſible, are very imperfect, and ſome of the courſes of experiments are incomplete; but a conſiderable number of facts, which appear to me to be new and important, are ſufficiently aſcertained; and I am willing to hope, that when philoſophers in general are apprized of them, ſome perſons may be able to purſue them to more advantage than myſelf. I therefore think it my duty to give this Society an account of the progreſs I have been able to make; and I ſhall not fail to communicate any farther lights that may occur to me, whenever I reſume theſe inquiries.

In writing upon this ſubject, I find myſelf at a loſs for proper terms, by which to diſtinguiſh the different kinds of air. Thoſe which have hitherto obtained are by no means ſufficiently characteriſtic, or diſtinct. The terms in common uſe are, fixed air, mephitic, and inflammable. The laſt, indeed, ſufficiently characterizes and diſtinguiſhes that kind of air which takes fire, and explodes on the approach of flame; but it might have been termed fixed with as much propriety as that to which Dr. Black and others have given that denomination, ſince it is originally part of ſome ſolid ſubſtance, and exiſts in an unelaſtic ſtate, and therefore may be alſo called factitious. The term mephitic is equally applicable to what is called fixed air, to that which is inflammable, and to many other kinds; ſince they are equally noxious, when breathed by animals. Rather, however, than to introduce new terms, or change the ſignification of old ones, I ſhall uſe the term fixed air, in the ſenſe in which it is now commonly uſed, and diſtinguiſh the other kinds by their properties, or ſome other periphraſis. I ſhall be under a neceſſity, however, of giving a name to one ſpecies of air, to which no name was given before.

Of FIXED AIR.

Fixed air is that which is expelled by heat from lime, and other calcareous ſubſtances, and, when deprived of which, they become quick-lime. It is alſo contained in alkaline ſalts, and is generated in great quantities from fermenting vegetables; and being united with water, gives it the principal properties of Pyrmont-water. This kind of air is alſo well known to be fatal to animals; and Dr. Macbride has demonſtrated, that it checks or prevents putrefaction.

Living for ſome time in the neighbourhood of a public brewery, I was induced to make a few experiments on this kind of air, there being always a large body of it, ready formed, upon the ſurface of the fermenting liquor, generally about nine inches or a foot in depth, within which any kind of ſubſtance may be very conveniently placed; and though it muſt be continually mixing with the common air, and is far from being perfectly pure, yet there is a conſtant ſupply from the fermenting liquor, and it is pure enough for many purpoſes.

A perſon, who is quite a ſtranger to the properties of this kind of air, would be agreeably amuſed with extinguiſhing lighted candles, or chips of wood in it, as it lies upon the ſurface of the fermenting liquor; for the ſmoke readily unites with this kind of air, probably by means of the water which it contains; ſo that very little or none of the ſmoke will eſcape into the open air, which is incumbent upon it. It is remarkable, that the upper ſurface of this ſmoke, floating in the fixed air, is ſmooth, and well defined; whereas the lower ſurface is exceedingly ragged, ſeveral parts hanging down to a conſiderable diſtance within the body of the fixed air, and ſometimes in the form of balls, connected to the upper ſtratum by ſlender threads, as if they were ſuſpended. The ſmoke is alſo apt to form itſelf into broad flakes, parallel to the ſurface of the liquor, and at different diſtances from it, exactly like clouds. Theſe appearances will ſometimes continue above an hour, with very little variation. When this fixed air is very ſtrong, the ſmoke of a ſmall quantity of gunpowder fired in it will be wholly retained by it, no part eſcaping into the common air.

Making an agitation in this air, the ſurface of it, which ſtill continues to be exactly defined, is thrown into the form of waves, which it is very amuſing to look upon; and if, by this agitation, any of the fixed

air

air be thrown over the ſide of the veſſel, the ſmoke, which is mixed with it, will fall to the ground, as if it was ſo much water, the fixed air being heavier than common air.

The red part of burning wood was extinguiſhed in this air, but I could not perceive that a red-hot poker was ſooner cooled in it.

Fixed air does not inſtantly mix with common air. Indeed, if it did, it could not be caught upon the fermenting liquor; for a candle put under a large receiver, and immediately plunged very deep below the ſurface of the fixed air, will burn ſome time. But veſſels with the ſmalleſt orifices, hanging with their mouths downwards in the fixed air, will in time have the common air, which they contain, perfectly mixed with it. When the fermenting liquor is contained in veſſels cloſe covered up, the fixed air is rendered much ſtronger, and then it readily affects the common air which is contiguous to it; ſo that, upon removing the cover, candles held at a conſiderable diſtance above the ſurface will inſtantly go out. I have been told by the workmen, that this will ſometimes be the caſe, when the candles are held more than half a yard above the mouth of the veſſel.

Fixed air unites with the ſmoke of reſin, ſulphur, and other electrical ſubſtances, as well as with the vapour of water; and yet, by holding the wire of a charged phial among theſe fumes, I could not make any electrical atmoſphere, which ſurprized me a good deal, as there was a large body of this ſmoke, and it was ſo confined, that it could not eſcape me. I alſo held ſome oil of vitriol in a glaſs veſſel within the

the fixed air, and by plunging a piece of red hot glaſs into it, raiſed a copious and thick fume. This floated upon the ſurface of the fixed air like other fumes, and continued as long.

Conſidering the near affinity between water and fixed air, I concluded that if a quantity of water was placed near the yeaſt of the fermenting liquor, it could not fail to imbibe that air, and thereby acquire the principal properties of Pyrmont, and other medicinal mineral waters. Accordingly, I found, that when the ſurface of the water was conſiderable, it always acquired the pleaſant acidulous taſte that Pyrmont water has. The readieſt way of impregnating water with this virtue, in theſe circumſtances, is to take two veſſels, and to keep pouring the water from one into the other, when they are both of them held as near the yeaſt as poſſible; for by this means a great quantity of ſurface is expoſed to the air, and the ſurface is alſo continually changing. In this manner, I have ſometimes, in the ſpace of two or three minutes, made a glaſs of exceedingly pleaſant ſparkling water, which could hardly be diſtinguiſhed from very good Pyrmont.

But the moſt effectual way of impregnating water with fixed air is to put the veſſels which contain the water into glaſs jars, filled with the pureſt fixed air, made by the ſolution of chalk in diluted oil of vitriol, ſtanding in quickſilver. In this manner I have, in about two days, made a quantity of water to imbibe more than an equal bulk of fixed air, ſo that, according to Dr. Brownrigg's experiments, it muſt have been much ſtronger than the beſt imported Pyrmont; for though he made his experiments at the ſpring head,

head, he never found that it contained quite ſo much as half its bulk of this air. If a ſufficient quantity of quickſilver cannot be procured, oil may be uſed with ſufficient advantage, for this purpoſe, as it imbibes the fixed air very ſlowly. Fixed air may be kept in veſſels ſtanding in water for a long time, if they be ſeparated by a partition of oil, about half an inch thick. Pyrmont water made in theſe circumſtances, is little or nothing inferior to that which has ſtood in quickſilver.

The *readieſt* method of preparing this water for uſe is to agitate it ſtrongly with its whole ſurface expoſed to the fixed air. By this means alſo, more than an equal bulk of air may be communicated to a large quantity of water in the ſpace of a few minutes. Eaſy directions for doing this I have publiſhed in a ſmall pamphlet, deſigned originally for the uſe of ſeamen in long voyages, on the preſumption that it might be of uſe for preventing or curing the ſea ſcurvy, equally with wort, which was recommended by Dr. Macbride for this purpoſe, on no other account than its property of generating fixed air, by its fermentation in the ſtomach.

Water thus impregnated with fixed air readily diſſolves iron, as Mr. Lane has diſcovered; ſo that if a quantity of iron filings be put to it, it preſently becomes a ſtrong chalybeate, and of the mildeſt and moſt agreeable kind.

I have recommended the uſe of chalk and oil of vitriol as the cheapeſt, and, upon the whole, the beſt materials for this purpoſe; and whereas ſome perſons had ſuſpected that a quantity of the oil of vitriol was rendered volatile by this proceſs, I examined it

by

by all the chemical methods that are in uſe; but could not find that water thus impregnated contained the leaſt perceivable quantity of the acid.

Mr. Hey, indeed, who aſſiſted me in this examination, found that diſtilled water, impregnated with fixed air, did not mix ſo readily with ſoap as the diſtilled water itſelf; but this was alſo the caſe when the fixed air had paſſed through a long glaſs tube filled with alkaline ſalts, which, it may be ſuppoſed, would have imbibed any of the oil of vitriol that might have been contained in that air *.

It is not improbable but that fixed air itſelf may be of the nature of an acid, though of a weak and peculiar ſort. Mr. Bergman of Upſal, who honoured me with a letter upon the ſubject, calls it the aërial acid, and, among other experiments to prove it to be an acid, he ſays that it changes the blue juice of tourneſole into red.

The heat of boiling water will expell all the fixed air, if a phial containing the impregnated water be held in it; but it will often require above half an hour to do it completely.

Dr. Percival, who is particularly attentive to every improvement in the medical art, and who has thought ſo well of this impregnation as to preſcribe it in ſeveral caſes, informs me that it ſeems to be much ſtronger, and ſparkles more, like the true Pyrmont water, after it has been kept ſome time. This circumſtance, however, ſhews that, in time, the fixed air is more eaſily diſengaged from the water, and

* An account of Mr. Hey's experiments will be found in the Appendix to theſe papers.

though, in this ſtate, it may affect the taſte more ſenſibly, it cannot be of ſo much uſe in the ſtomach and bowels, as when the air is more firmly retained by the water, though, in conſequence of it, it be leſs ſenſible to the taſte.

By the proceſs deſcribed in my pamphlet, fixed air may be readily incorporated with wine, beer, and almoſt any other liquor whatever; and when beer, wine, or cyder, is become flat or dead (which is the conſequence of the eſcape of the fixed air they contained) they may be revived by this means; but the delicate and agreeable flavour, or acidulous taſte, communicated by fixed air, and which is very manifeſt in water, can hardly be perceived in wine, or any liquors which have much taſte of their own.

I ſhould think that there can be no doubt, but that water thus impregnated with fixed air muſt have all the medicinal virtues of genuine Pyrmont water; ſince theſe depend upon the fixed air it contains. If the genuine Pyrmont water derives any advantage from its being a natural chalybeate, this may alſo be obtained by providing a common chalybeate water, and uſing it in theſe proceſſes, inſtead of common water.

Having ſucceeded ſo well with this artificial Pyrmont water, I imagined that it might be poſſible to give ice the ſame virtue, eſpecially as cold is known to promote the abſorption of fixed air by water; but in this I found myſelf quite miſtaken. I put ſeveral pieces of ice into a quantity of fixed air, confined by quickſilver, but no part of the air was abſorbed in two days and two nights; but upon bringing it into a place where the ice melted, the air was abſorbed as uſual. I then took a quantity of ſtrong artificial Pyrmont water, and, putting it into a thin glaſs phial, I ſet it in a pot that was filled with ſnow and ſalt. This mixture inſtantly freezing the water that was contiguous to the ſides of the glaſs, the air was diſcharged plentifully, ſo that I catched a conſiderable quantity, in a bladder tied to the mouth of the phial. I alſo took two quantities of the ſame Pyrmont water, and placed one of them where it might freeze, keeping the other in a cold place, but where it would not freeze. This retained its acidulous taſte, though the phial which contained it was not corked; whereas the other, being brought into the ſame place, where the ice melted very ſlowly, had at the ſame time the taſte of common water only. That quantity of water which had been frozen by the mixture of ſnow and ſalt, was almoſt as much like ſnow as ice, ſuch a quantity of air bubbles were contained in it, by which it was prodigiouſly increaſed in bulk.

The preſſure of the atmoſphere aſſiſts very conſiderably in keeping fixed air confined in water; for in an exhauſted receiver, Pyrmont water will abſolutely boil, by the copious diſcharge of its air. This is alſo the reaſon why beer and ale froth ſo much *in vacuo*. I do not doubt, therefore, but that, by the help of a condenſing engine, water might be much more highly impregnated with the virtues of the Pyrmont ſpring, and it would not be difficult to contrive a method of doing it.

The manner in which I made ſeveral experiments to aſcertain the abſorption of fixed air by different fluid ſubſtances was to put the liquid into a diſh, and holding it within the body of the fixed air at the brewery, to ſet a glaſs veſſel into it, with its mouth inverted. This glaſs being neceſſarily filled with the fixed air, the liquor would riſe into it when they were both taken into the common air, if the fixed air was abſorbed at all.

Making uſe of ether in this manner, there was a conſtant bubbling from under the glaſs, occaſioned by this fluid eaſily riſing in vapour, ſo that I could not, in this method, determine whether it imbibed the air or not. I concluded, however, that they did incorporate, from a very diſagreeable circumſtance, which made me deſiſt from making any more experiments of the kind. For all the beer, over which this experiment was made, contracted a peculiar taſte, the fixed air impregnated with the ether being, I ſuppoſe, again abſorbed by the beer. I have alſo obſerved, that water which remained a long time within this air has ſometimes acquired a very diſagreeable taſte. At one time it was like tar-water. How this was acquired, I was very deſirous of making ſome experiments to aſcertain, but I was diſcouraged by the fear of injuring the fermenting liquor. It could not come from the fixed air only.

Having imagined that fixed air coagulated the blood in the lungs of animals, and thereby cauſed inſtant death; I ſuffocated a cat in this kind of air, and examining the lungs preſently after, found them collapſed and white, having little or no blood in them.

In order to try the effect of this air upon the blood itſelf, I took a quantity from a fowl juſt killed, and divided it into two parts, holding one of them within the

the fixed air, and the other in the common air, and obſerved that the former was coagulated much ſooner than the latter. This I could wiſh to have tried again.

Inſects and animals which breathe very little are ſtifled in fixed air, but are not ſoon quite killed in it. Butterflies, and flies of other kinds, will generally become torpid, and ſeemingly dead, after being held a few minutes over the fermenting liquor; but they revive again after being brought into the freſh air. But there are very great varieties with reſpect to the time in which different kinds of flies will either become torpid in the fixed air, or die in it. A large ſtrong frog was much ſwelled, and ſeemed to be nearly dead, after being held about ſix minutes over the fermenting liquor; but it recovered upon being brought into the common air. A ſnail treated in the ſame manner died preſently.

Fixed air is preſently fatal to vegetable life. At leaſt ſprigs of mint, growing in water, and placed over the fermenting liquor, will often become quite dead in one day, or even in a leſs ſpace of time; nor do they recover when they are afterwards brought into the common air. I am told, however, that ſome other plants are much more hardy in this reſpect.

A red roſe, freſh gathered, loſt its redneſs, and became of a purple colour, after being held over the fermenting liquor about twenty-four hours; but the tips of each leaf were much more affected than the reſt of it. Another red roſe turned perfectly white in this ſituation; but various other flowers, of different colours, were very little affected. Theſe experiments

riments were not repeated, as I wiſh they might be done, in pure fixed air, extracted from chalk by means of oil of vitriol.

For every purpoſe, in which it was neceſſary that the fixed air ſhould be as unmixed as poſſible, I generally made it by pouring oil of vitriol upon chalk and water, catching it in a bladder, faſtened to the neck of the phial, in which they were contained, taking care to preſs out all the common air, and alſo the firſt, and ſometimes the ſecond, produce of fixed air; and alſo, by agitation, making it as quickly as I poſſibly could. At other times, I made it paſs from the phial in which it was generated through a glaſs tube, without the intervention of any bladder, which, as I found by experience, will not long make a ſufficient ſeparation between ſeveral kinds of air and common air.

I had once thought that the readieſt method of procuring fixed air, and in ſufficient purity, would be by the ſimple proceſs of burning chalk, or pounded lime-ſtone in a gun-barrel, making it paſs through the ſtem of a tobacco-pipe, or a glaſs tube carefully luted to the orifice of it; and in this manner I find that air is produced in great plenty; but, upon examining it, I found, to my very great ſurprize, that little more than one half of it was fixed air, capable of being abſorbed by water; and that the reſt was inflammable, ſometimes very weakly, but ſometimes pretty highly ſo. Whence this inflammability proceeds, I am not able to determine, the lime or chalk not being ſuppoſed to contain any other than fixed air. I conjecture, however, that it muſt proceed from the iron, and the ſeparation of it

from

from the calx may be promoted by that ſmall quantity of oil of vitriol, which I am informed is contained in chalk, if not in lime-ſtone alſo. But it is an objection to this hypotheſis, that the inflammable air produced in this manner burns blue, and not at all like that which is produced from iron, or any other metal, by means of an acid. It has alſo the ſmell of that kind of inflammable air which is produced from vegetable ſubſtances. Beſides, oil of vitriol without water, will not diſſolve iron; nor can inflammable air be got from it, unleſs the acid be conſiderably diluted; and when I mixed brimſtone with the chalk, neither the quality nor the quantity of the air was changed by it. Indeed no air, or permanently elaſtic vapour, can be got from brimſtone, or any oil.

In the method in which I generally made the fixed air, and indeed always, unleſs the contrary be particularly mentioned, *viz.* by diluted oil of vitriol and chalk, I found by experiment that it was as pure as Mr. Cavendiſh made it. For after it had paſſed through a large body of water in ſmall bubbles, ſtill $\frac{1}{50}$ or $\frac{1}{60}$ part only was not abſorbed by water. In order to try this as expeditiouſly as poſſible, I kept pouring the air from one glaſs veſſel into another, immerſed in a quantity of cold water, in which manner I found by experience, that almoſt any quantity may be reduced as far as poſſible in little more than a quarter of an hour.

At the ſame time that I was trying the purity of my fixed air, I had the curioſity to endeavour to aſcertain whether that part of it which is not miſſcible in water, be equally diffuſed through the whole

maſs;

maſs; and, for this purpoſe, I divided a quantity of about a gallon into three parts, the firſt conſiſting of that which was uppermoſt, and the laſt of that which was the loweſt, contiguous to the water; but all theſe parts were reduced in about an equal proportion, by paſſing through the water, ſo that the whole maſs had been of an uniform compoſition. This I have alſo found to be the caſe with ſeveral kinds of air, which will not properly incorporate.

A mouſe will live very well, though a candle will not burn, in the reſiduum of the pureſt fixed air that I can make; and I once made a very large quantity for the ſole purpoſe of this experiment. This, therefore, ſeems to be one inſtance of the generation of genuine common air, though vitiated in ſome degree. It is alſo another proof of the reſiduum of fixed air being, in part at leaſt, common air, that it becomes turbid, and is diminiſhed by the mixture of nitrous air, as will be explained hereafter.

That fixed air only wants ſome addition to make it permanent, and immiſcible with water, if not, in all reſpects, common air, I have been led to conclude, from ſeveral attempts which I once made to mix it with air, in which a quantity of iron filings and brimſtone, made into a paſte with water, had ſtood; for, in ſeveral mixtures of this kind, I imagined that not much more than half of the fixed air could be imbibed by water; but, not being able to repeat the experiment, I conclude that I either deceived myſelf in it, or that I overlooked ſome circumſtance on which the ſucceſs of it depended.

Theſe experiments, however, whether they were fallacious or otherwiſe, induced me to try whether any

any alteration would be made in the conſtitution of fixed air, by this mixture of iron filings and brimſtone. I therefore put a mixture of this kind into a quantity of as pure fixed air as I could make, and confined the whole in quickſilver, leſt the water ſhould abſorbe it before the effects of the mixture could take place. The conſequence was, that the fixed air was diminiſhed, and the quickſilver roſe in the veſſel, till about the fifth part was occupied by it; and, as near as I could judge, the proceſs went on, in all reſpects, as if the air in the inſide had been common air.

What is moſt remarkable, in the reſult of this experiment, is, that the fixed air, into which this mixture had been put, and which had been in part diminiſhed by it, was in part alſo rendered inſoluble in water by this means. I made this experiment four times, with the greateſt care, and obſerved, that in two of them about one ſixth, and in the other two about one fourteenth, of the original quantity, was ſuch as could not be abſorbed by water, but continued permanently elaſtic. Leſt I ſhould have made any miſtake with reſpect to the purity of the fixed air, the laſt time that I made the experiment, I ſet part of the fixed air, which I made uſe of, in a ſeparate veſſel, and found it to be exceedingly pure, ſo as to be almoſt wholly abſorbed by water; whereas the other part, to which I had put the mixture, was far from being ſo.

In one of theſe caſes, in which fixed air was made immiſcible with water, it appeared to be not very noxious to animals; but in another caſe, a mouſe died in it pretty ſoon.

As the iron is reduced to a calx by this proceſs, I once concluded, that it is phlogiſton that fixed air wants, to make it common air; and, for any thing I yet know, this may be the caſe, though I am ignorant of the method of combining them; and when I calcined a quantity of lead in fixed air, in the manner which will be deſcribed hereafter, it did not ſeem to have been leſs ſoluble in water than it was before.

II.

On Air in which a candle, or brimstone, has burned out.

It is well known that flame cannot ſubſiſt long without change of air, ſo that the common air is neceſſary to it, except in the caſe of ſubſtances, into the compoſition of which nitre enters; for theſe will burn *in vacuo*, in fixed air, and even under water, as is evident in ſome rockets, which are made for this purpoſe. The quantity of air which even a ſmall flame requires to keep it burning is prodigious. It is generally ſaid, that an ordinary candle conſumes, as it is called, about a gallon in a minute. Conſidering this amazing conſumption of air, by fires of all kinds, volcano's, &c. it becomes a great object of philoſophical inquiry, to aſcertain what change is made in the conſtitution of the air by flame, and to diſcover what proviſion there is in nature for remedying the injury which the atmoſphere receives by this means. Some of the following experiments will, perhaps, be thought to throw a little light upon the ſubject.

The

The diminution of the quantity of air in which a candle, or brimſtone, has burned out, is various; but I imagine that, at a medium, it may be about one fifteenth, or one ſixteenth, of the whole; about one third as much as by animals breathing it as long as they can, by animal or vegetable ſubſtances putrifying in it, by the calcination of metals, or by a mixture of ſteel filings and pounded brimſtone ſtanding in it.

I have ſometimes thought, that flame diſpoſes the common air to depoſit the fixed air it contains; for if any lime-water be expoſed to it, it immediately becomes turbid. This is the caſe, when wax candles, tallow candles, chips of wood, ſpirit of wine, æther, and every other ſubſtance which I have yet tried, except brimſtone, is burned in a cloſe glaſs veſſel, ſtanding in lime-water. This precipitation of fixed air (if this be the caſe) may be owing to ſomething emitted from the burning bodies, which has a ſtronger affinity with the other conſtituent parts of the atmoſphere.

If brimſtone be burned in the ſame circumſtances, the lime-water continues tranſparent, but ſtill there may have been the ſame precipitation of the fixed part of the air; but that, uniting with the lime and the vitriolic acid, it forms a ſelenetic ſalt, which is ſoluble in water. Having evaporated a quantity of water thus impregnated, by burning brimſtone a great number of times over it, a whitiſh powder remained, which had an acid taſte; but repeating the experiment with a quicker evaporation, the powder had no acidity, but was very much like chalk. The burning of brimſtone but once over a quantity of lime-water, will affect it in such a manner, that breathing into it will not make it turbid, which otherwise it always presently does.

Dr. Hales supposed, that by burning brimstone repeatedly in the same quantity of air, the diminution would continue without end. But this I have frequently tried, and not found to be the case. Indeed, when the ignition has been imperfect in the first instance, a second firing of the same substance will increase the effect of the first, &c. but this progress soon ceases. In many cases of the diminution of air, the effect is not immediately apparent, even when it stands in water; for sometimes the bulk of air will not be much reduced, till it has passed several times through a quantity of water, which has thereby a better opportunity of absorbing that fluid part of the air, which had not been perfectly detached from the rest. I have sometimes found a very great reduction of a mass of air, in consequence of passing but once thorough cold water. If the air has stood in quicksilver, the diminution is generally inconsiderable, till it has undergone this operation, there not being any substance exposed to the air that could absorb any part of it.

I could not find any considerable alteration in the specific gravity of the air, in which candles, or brimstone, had burned out. I am satisfied, however, that it is not heavier than common air, which must have been manifest, if so great a diminution of the quantity had been owing, as Dr. Hales and others supposed, to the elasticity of the whole mass being impaired. After making several trials for this purpose, I concluded that air, thus diminished in bulk, is rather lighter than common air, which favours the suppofition of the fixed, or heavier part of the common air, having been precipitated.

An animal will live nearly, if not quite as long, in air in which candles have burned out, as in common air. This fact furprized me very greatly, having imagined that what is called the confumption of air by flame, or refpiration, to have been of the fame nature; but I have fince found, that this fact has been obferved by many perfons, and even fo early as by Mr. Boyle. I have alfo obferved, that air in which brimftone has burned, is not in the leaft injurious to animals, after the fumes, which at firft make it very cloudy, have intirely fubfided.

Having read, in the Memoirs of the Society at Turin, Vol. I. p. 41. that air in which candles had burned out was perfectly reftored, fo that other candles would burn in it again as well as ever, after having been expofed to a confiderable degree of cold, and likewife after having been compreffed in bladders (for the cold had been fuppofed to have produced this effect by nothing but condenfation): I repeated thefe experiments, and did, indeed, find, that, when I compreffed the air in bladders, as the Count de Saluce, who made the obfervation, had done, the experiment fucceeded: but having had fufficient reafon to diftruft bladders, I compreffed the air in a glafs veffel ftanding in water; and then I found, that this procefs is altogether ineffectual for the purpofe. I kept the air compreffed much more, and much longer, than he had done, but without producing any alteration in it. I alfo find, that a greater degree of cold than that which he applied, and

of

of longer continuance, did by no means reſtore this kind of air: for when I have expoſed the phials which contained it a whole night, in which the froſt was very intenſe; and alſo when I kept it ſurrounded with a mixture of ſnow and ſalt, I found it, in all reſpects, the ſame as before.

It is alſo advanced, in the ſame Memoir, p. 41. that heat only, as the reverſe of cold, renders air unfit for candles burning in it. But I repeated the experiment of the Count for that purpoſe, without finding any ſuch effect from it. I alſo remember that, many years ago, I filled an exhauſted receiver with air, that had paſſed through a glaſs tube made red-hot, and found that a candle would burn in it perfectly well. Alſo, rarefaction by the air-pump does not injure air in the leaſt degree.

Though this experiment failed, I flatter myſelf that I have accidentally hit upon a method of reſtoring air which has been injured by the burning of candles, and that I have diſcovered at leaſt one of the reſtoratives which nature employs for this purpoſe. It is vegetation. In what manner this proceſs in nature operates, to produce ſo remarkable an effect, I do not pretend to have diſcovered; but a number of facts declare in favour of this hypotheſis. I ſhall introduce my account of them, by reciting ſome of the obſervations which I made on the growing of plants in confined air, which led to this diſcovery.

One might have imagined that, ſince common air is neceſſary to vegetable, as well as to animal life, both plants and animals had affected it in the ſame manner, and I own I had that expectation, when

when I firſt put a ſprig of mint into a glaſs-jar, ſtanding inverted in a veſſel of water; but when it had continued growing there for ſome months, I found that the air would neither extinguiſh a candle, nor was it at all inconvenient to a mouſe, which I put into it.

The plant was not affected any otherwiſe than was the neceſſary conſequence of its confined ſituation; for plants growing in ſeveral other kinds of air, were all affected in the very ſame manner. Every ſucceſſion of leaves was more diminiſhed in ſize than the preceding, till, at length, they came to be no bigger than the heads of pins. The root decayed, and the ſtalk alſo, beginning from the root; and yet the plant continued to grow upwards, drawing its nouriſhment through a black and rotten ſtem. In the third or fourth ſet of leaves, long hairy filaments grew from the inſertion of each leaf, and ſometimes from the body of the ſtem, ſhooting out as far as the veſſel in which it grew would permit, which, in my experiments, was about two inches. In this manner a ſprig of mint lived, the old ſtem decaying, and new ones ſhooting up in its place, but leſs and leſs continually, all the ſummer ſeaſon.

In repeating this experiment, care muſt be taken to draw away all the dead leaves from about the plant, leſt they ſhould putrefy, and affect the air. I have found that a freſh cabbage leaf, put under a glaſs veſſel filled with common air, for the ſpace of one night only, has ſo far affected the air, that a candle would not burn in it the next morning, and yet the leaf had not acquired any ſmell of putrefaction.

Finding that candles burn very well in air in which plants had grown a long time, and having had ſome reaſon to think, that there was ſomething attending vegetation, which reſtored air that had been injured by reſpiration, I thought it was poſſible that the ſame proceſs might alſo reſtore the air that had been injured by the burning of candles.

Accordingly, on the 17th of Auguſt, 1771, I put a ſprig of mint into a quantity of air, in which a wax candle had burned out, and found that, on the 27th of the ſame month, another candle burned perfectly well in it. This experiment I repeated, without the leaſt variation in the event, not leſs than eight or ten times in the remainder of the ſummer. Several times I divided the quantity of air in which the candle had burned out, into two parts, and putting the plant into one of them, left the other in the ſame expoſure, contained, alſo, in a glaſs veſſel immerſed in water, but without any plant; and never failed to find, that a candle would burn in the former, but not in the latter. I generally found that five or ſix days were ſufficient to reſtore this air, when the plant was in its vigour; whereas I have kept this kind of air in glaſs veſſels, immerſed in water many months, without being able to perceive that the leaſt alteration had been made in it. I have alſo tried a great variety of experiments upon it, as by condenſing, rarefying, expoſing to the light and heat, &c. and throwing into it the effluvia of many different ſubſtances, but without any effect.

Experiments made in the year 1772, abundantly confirmed my concluſion concerning the reſtoration of air, in which candles had burned out by plants growing

growing in it. The firſt of theſe experiments was made in the month of May; and they were frequently repeated in that and the two following months, without a ſingle failure.

For this purpoſe I uſed the flames of different ſubſtances, though I generally uſed wax or tallow candles. On the 24th of June the experiment ſucceeded perfectly well with air in which ſpirit of wine had burned out, and on the 27th of the ſame month it ſucceeded equally well with air in which brimſtone matches had burned out, an effect of which I had deſpaired the preceding year.

This reſtoration of air I found depended upon the vegetating ſtate of the plant; for though I kept a great number of the freſh leaves of mint in a ſmall quantity of air in which candles had burned out, and changed them frequently, for a long ſpace of time, I could perceive no melioration in the ſtate of the air.

This remarkable effect does not depend upon any thing peculiar to mint, which was the plant that I always made uſe of till July 1772; for on the 16th of that month, I found a quantity of this kind of air to be perfectly reſtored by ſprigs of balm, which had grown in it from the 7th of the ſame month.

That this reſtoration of air was not owing to any aromatic effluvia of theſe two plants, not only appeared by the eſſential oil of mint having no ſenſible effect of this kind; but from the equally complete reſtoration of this vitiated air by the plant called groundſel, which is uſually ranked among the weeds, and has an offenſive ſmell. This was the reſult of an experiment made the 16th of July, when the

 plant

plant had been growing in the burned air from the 8th of the ſame month. Beſides, the plant which I have found to be the moſt effectual of any that I have tried for this purpoſe is ſpinach, which is of quick growth, but will ſeldom thrive long in water. One jar of burned air was perfectly reſtored by this plant in four days, and another in two days. This laſt was obſerved on the 22d of July. In general this effect may be preſumed to have taken place in much leſs time than I have mentioned; becauſe I never choſe to make a trial of the air, till I was pretty ſure, from preceding obſervations, that the event which I had expected muſt have taken place, if it would ſucceed at all; leſt, returning back that part of the air on which I made the trial, and which would thereby neceſſarily receive a ſmall mixture of common air, the experiment might not be judged to be quite fair; though I myſelf might be ſufficiently ſatisfied with reſpect to the allowance that was to be made for that ſmall imperfection.

III.

Of inflammable Air.

I have generally made inflammable air in the manner deſcribed by Mr. Cavendiſh, in the Philoſophical Tranſactions, from iron, zinc, or tin; but chiefly from the two former metals, on account of the proceſs being the leaſt troubleſome: but when I extracted it from vegetable or animal ſubſtances, or from coals, I put them into a gun barrel, to the orifice of which I luted a glaſs tube, or the ſtem of

a to-a tobacco pipe, and to the end of this I tied a flaccid bladder, in order to catch the generated air.

There is not, I believe, any vegetable or animal ſubſtance whatever, nor any mineral ſubſtance, that is inflammable, but what will yield great plenty of inflammable air, when they are treated in this manner, and urged with a ſtrong heat; but, in order to get the moſt air, the heat muſt be applied as ſuddenly, and as vehemently, as poſſible. For, notwithſtanding the ſame care be taken in luting, and in every other reſpect, ſix or even ten times more air may be got by a ſudden heat than by a ſlow one, though the heat that is laſt applied be as intenſe as that which was applied ſuddenly. A bit of dry oak, weighing about twelve grains, will generally yield about a ſheep's bladder full of inflammable air with a briſk heat, when it will only give about two or three ounce meaſures if the ſame heat be applied to it very gradually. To what this difference is owing, I cannot tell.

Inflammable air, when it is made by a quick proceſs, has a very ſtrong and offenſive ſmell, from whatever ſubſtance it be generated; but this ſmell is of three different kinds, according as the air is extracted from mineral, vegetable, or animal ſubſtances. The laſt is exceedingly fetid; and it makes no difference, whether it be extracted from a bone, or even an old and dry tooth, or from ſoft muſcular fleſh, or any other part of the animal. The burning of any ſubſtance occaſions the ſame ſmell: for the groſs fume which ariſes from them, before they flame, is the inflammable air they contain, which is expelled by heat, and then readily ignited. The ſmell of inflammable air is the very ſame, as far as I am able to perceive, from whatever ſubſtance of the ſame kingdom it be extracted. Thus it makes no difference whether it be got from iron, zinc, or tin, from any kind of wood, or, as was obſerved before, from any part of an animal.

If a quantity of inflammable air be contained in a glaſs veſſel ſtanding in water, and have been generated very faſt, it will ſmell even through the water, and this water will alſo ſoon become covered with a thin film, aſſuming all the different colours. If the inflammable air have been generated from iron, this matter will appear to be a red okre, or the earth of iron, as I have found by collecting a conſiderable quantity of it; and if it have been generated from zinc, it is a whitiſh ſubſtance, which I ſuppoſe to be the calx of the metal. It likewiſe ſettles to the bottom of the veſſel, and when the water is ſtirred, it has very much the appearance of wool. When water is once impregnated in this manner, it will continue to yield this ſcum for a conſiderable time after the air is removed from it. This I have often obſerved with reſpect to iron.

Inflammable air, made by a violent effervеſcence, I have obſerved to be much more inflammable than that which is made by a weak effervеſcence, whether the water or the oil of vitriol prevailed in the mixture. Alſo the offenſive ſmell was much ſtronger in the former caſe than in the latter. The greater degree of inflammability appeared by the greater number of ſucceſſive exploſions, when a candle was preſented to the neck of a phial filled with it. It is poſſible, however, that this diminution of inflammability

flammability may, in ſome meaſure, ariſe from the air continuing ſo much longer in the bladder when it is made very ſlowly; though I think the difference is too great for this cauſe to have produced the whole of it. It may, perhaps, deſerve to be tried by a different proceſs, without a bladder.

Inflammable air is not thought to be miſcible with water, and when kept many months, ſeems, in general, to be as inflammable as ever. Indeed, when it is extracted from vegetable or animal ſubſtances, a part of it will be imbibed by the water in which it ſtands; but it may be preſumed, that in this caſe, there was a mixture of fixed air extracted from the ſubſtance along with it. I have indiſputable evidence, however, that inflammable air, ſtanding long in water, has actually loſt all its inflammability, and even come to extinguiſh flame much more than that air in which candles have burned out. After this change it appears to be greatly diminiſhed in quantity, and it ſtill continues to kill animals the moment they are put into it.

This very remarkable fact firſt occurred to my obſervation on the twenty-fifth of May 1771, when I was examining a quantity of inflammable air, which had been made from zinc, near three years before. Upon this, I immediately ſet by a common quart bottle filled with inflammable air from iron, and another equal quantity from zinc; and examining them in the beginning of December following, that from the iron was reduced near one half in quantity, if I be not greatly miſtaken; for I found the bottle half full of water, and I am pretty clear that it was full of air when it was ſet by. That which had been

been produced from zinc was not altered, and filled the bottle as at firſt.

Another inſtance of this kind occurred to my obſervation on the 19th of June 1772, when a quantity of air, half of which had been inflammable air from zinc, and half air in which mice had died, and which had been put together the 30th of July 1771, appeared not to be in the leaſt inflammable, but extinguiſhed flame, as much as any kind of air that I had ever tried. I think that, in all, I have had four inſtances of inflammable air loſing its inflammability, while it ſtood in water.

Though air tainted with putrefaction extinguiſhes flame, I have not found that animals or vegetables putrefying in inflammable air render it leſs inflammable. But one quantity of inflammable air, which I had ſet by in May 1771, along with the others above mentioned, had had ſome putrid fleſh in it; and this air had loſt its inflammability, when it was examined at the ſame time with the other in the December following. The bottle in which this air had been kept, ſmelled exactly like very ſtrong Harrowgate water. I do not think that any perſon could have diſtinguiſhed them.

I have made plants grow for ſeveral months in inflammable air made from zinc, and alſo from oak; but, though the plants grew pretty well, the air ſtill continued inflammable. The former, indeed, was not ſo highly inflammable as when it was freſh made, but the latter was quite as much ſo; and the diminution of inflammability in the former caſe, I attribute to ſome other cauſe than the growth of the plant.

No

No kind of air, on which I have yet made the experiment, will conduct electricity; but the colour of a ſpark is remarkably different in ſome different kinds of air, which ſeems to ſhew that they are not equally good non-conductors. In fixed air, the electric ſpark is exceedingly white; but in inflammable air it is of a purple, or red colour. Now, ſince the moſt vigorous ſparks are always the whiteſt, and, in other caſes, when the ſpark is red, there is reaſon to think that the electric matter paſſes with difficulty, and with leſs rapidity: it is poſſible that the inflammable air may contain particles which conduct electricity, though very imperfectly; and that the whiteneſs of the ſpark in the fixed air, may be owing to its meeting with no conducting particles at all. When an exploſion was made in a quantity of inflammable air, it was a little white in the center, but the edges of it were ſtill tinged with a beautiful purple. The degree of whiteneſs in this caſe was probably owing to the electric matter ruſhing with more violence in an exploſion than in a common ſpark.

Inflammable air kills animals as ſuddenly as fixed air, and, as far as can be perceived, in the ſame manner, throwing them into convulſions, and thereby occaſioning preſent death. I had imagined that, by animals dying in a quantity of inflammable air, it would in time become leſs noxious; but this did not appear to be the caſe; for I killed a great number of mice in a ſmall quantity of this air, which I kept ſeveral months for this purpoſe, without its being at all ſenſibly mended; the laſt, as well as the firſt mouſe, dying the moment it was put into it.

I once imagined that, ſince fixed and inflammable air are the reverſe of one another, in ſeveral remarkable properties, a mixture of them would make common air; and while I made the mixtures in bladders, I imagined that I had ſucceeded in my attempt; but I have ſince found that thin bladders do not ſufficiently prevent the air that is contained in them from mixing with the external air. Alſo corks will not ſufficiently confine different kinds of air, unleſs the phials in which they are confined be ſet with their mouths downwards, and a little water lie in the necks of them, which, indeed, is equivalent to the air ſtanding in veſſels immerſed in water. In this manner, however, I have kept different kinds of air for ſeveral years.

Whatever methods I took to promote the mixture of fixed and inflammable air, they were all ineffectual. I think it my duty, however, to recite the iſſue of an experiment or two of this kind, in which equal mixtures of theſe two kinds of air had ſtood near three years, as they ſeem to ſhew that they had in part affected one another, in that long ſpace of time. Theſe mixtures I examined April 27, 1771. One of them had ſtood in quickſilver, and the other in a corked phial, with a little water in it. On opening the latter in water, the water inſtantly ruſhed in, and filled almoſt half of the phial, and very little more was abſorbed afterwards. In this caſe the water in the phial had probably abſorbed a conſiderable part of the fixed air, ſo that the inflammable air was exceedingly rarefied; and yet the whole quantity that muſt have been rendered non-elaſtic was ten times more than the bulk of the water, and it has not

not been found that water can contain much more than its own bulk of fixed air. But in other cafes I have found the diminution of a quantity of air, and efpecially of fixed air, to be much greater than I could well account for by any kind of abforption.

The phial which had ftood immerfed in quickfilver had loft very little of its original quantity; and being now opened in water, and left there, along with a another phial, which was juft then filled, as this had been three years before, with air half inflammable and half fixed, I obferved that the quantity of both was diminifhed, by the abforption of the water, in the fame proportion.

Upon applying a candle to the mouths of the phials which had been kept three years, that which had ftood in quickfilver went off at one explofion, exactly as it would have done if there had been a mixture of common air, with the inflammable. As a good deal depends upon the apertures of the veffels in which the inflammable air is fixed, I mixed the two kinds of air in equal proportion in the fame phial, and after letting it ftand fome days in water, that the fixed air might be abforbed, I applied a candle to it; but it made ten or twelve explofions (ftopping the phial after each of them) before the inflammable matter was exhaufted.

The air which had been confined in the corked phial exploded in the very fame manner as an equal mixture of the two kinds of air in the fame phial, the experiment being made as foon as the fixed air was abforbed, as before; fo that, in this cafe, the two kinds of air did not feem to have affected one another at all.

 Con-

Conſidering inflammable air as air united to or loaded with phlogiſton, I expoſed to it ſeveral ſubſtances, which are ſaid to have a near affinity with phlogiſton, as oil of vitriol, and ſpirit of nitre (the former for above a month), but without making any ſenſible alteration in it.

I obſerved, however, that inflammable air, mixed with the fumes of ſmoaking ſpirit of nitre, goes off at one exploſion, exactly like a mixture of half common and half inflammable air. This I tried ſeveral times, by throwing the inflammable air into a phial full of ſpirit of nitre, with its mouth immerſed in a baſon containing ſome of the ſame ſpirit, and then applying the flame of a candle to the mouth of the phial, the moment that it was uncovered, after it had been taken out of the baſon. This remarkable effect I haſtily concluded to have ariſen from the inflammable air having been in part deprived of its inflammability, by means of the ſtronger affinity, which the ſpirit of nitre had with phlogiſton, and therefore I imagined that by letting them ſtand longer in contact, and eſpecially by agitating them ſtrongly together, I ſhould deprive the air of all its inflammability; but neither of theſe operations ſucceeded, for ſtill the air was only exploded at once, as before. And laſtly, when I paſſed a quantity of inflammable air, which had been mixed with the fumes of ſpirit of nitre, through a body of water, and received it in another veſſel, it appeared not to have undergone any change at all, for it went off in ſeveral ſucceſſive exploſions, like the pureſt inflammable air. The effect abovementioned muſt, therefore, have been owing to the fumes of the ſpirit of nitre ſupplying the

the place of common air for the purpoſe of ignition, which is analogous to other experiments with nitre.

Having had the curioſity, on the 25th of July 1772, to expoſe a great variety of different kinds of air to water out of which the air it contained had been boiled, without any particular view; the reſult was, in ſeveral reſpects, altogether unexpected, and led to a variety of new obſervations on the properties and affinities of ſeveral kinds of air with reſpect to water. Among the reſt three fourths of that which was inflammable was abſorbed by the water in about two days, and the remainder was inflammable, but weakly ſo.

Upon this, I began to agitate a quantity of ſtrong inflammable air in a glaſs jar, ſtanding in a pretty large trough of water, the ſurface of which was expoſed to the common air, and I found that when I had continued the operation about ten minutes, near one fourth of the quantity of air had diſappeared; and finding that the remainder made an effervеſcence with nitrous air, I concluded that it muſt have become fit for reſpiration, whereas this kind of air is, at the firſt, as noxious as any other kind whatever. To aſcertain this, I put a mouſe into a veſſel containing 2½ ounce meaſures of it, and obſerved that it lived in it twenty minutes, which is as long as a mouſe will generally live in the ſame quantity of common air. This mouſe was even taken out alive, and recovered very well. Still alſo the air in which it had breathed ſo long was inflammable, though very weakly ſo. I have even found it to be ſo when a mouſe has actually died in it.

Inflammable air thus diminiſhed by agitation in water, makes but one exploſion on the approach of a candle exactly like a mixture of inflammable air with common air.

From this experiment I concluded that, by continuing the ſame proceſs, I ſhould deprive inflammable air of all its inflammability, and this I found to be the caſe; for, after a longer agitation, it admitted a candle to burn in it, like common air, only more faintly; and indeed by the teſt of nitrous air it did not appear to be near ſo good as common air. Continuing the ſame proceſs ſtill farther, the air which had been moſt ſtrongly inflammable a little before, came to extinguiſh a candle, exactly like air in which a candle had burned out, nor could they be diſtinguiſhed by the teſt of nitrous air.

I found, by repeated trials, that it was difficult to catch the time in which inflammable air obtained from metals, in coming to extinguiſh flame, was in the ſtate of common air, ſo that the tranſition from the one to the other muſt be very ſhort. I readily, however, found this ſtate in a quantity of inflammable air extracted from oak, which air I had kept by me a year, and in which a plant had grown, though very poorly, for ſome part of the time. A quantity of this air, after being agitated in water till it was diminiſhed about one half, admitted a candle to burn in it exceedingly well, and was even hardly to be diſtinguiſhed from common air by the teſt of nitrous air.

I took ſome pains to aſcertain the quantity of diminution, in freſh made and very highly inflammable air from iron, at which it ceaſed to be inflammable,

mable, and, upon the whole, I concluded that it was ſo when it was diminiſhed a little more than one half: for a quantity which was diminiſhed exactly one half had ſomething inflammable in it, but in the ſlighteſt degree imaginable.

Finding that water would imbibe inflammable air, I endeavoured to impregnate water with it, by the ſame proceſs by which I had made water imbibe fixed air; but though I found that diſtilled water would imbibe about one fourteenth of its bulk of inflammable air, I could not perceive that the taſte of it was ſenſibly altered.

IV.

OF AIR INFECTED WITH ANIMAL RESPIRATION, OR PUTREFACTION.

That candles will burn only a certain time, is a fact not better known, than it is that animals can live only a certain time, in a given quantity of air; but the cauſe of the death of the animal is not better known than that of the extinction of flame in the ſame circumſtances; and when once any quantity of air has been rendered noxious by animals breathing in it as long as they could, I do not know that any methods have been diſcovered of rendering it fit for breathing again. It is evident, however, that there muſt be ſome proviſion in nature for this purpoſe, as well as for that of rendering the air fit for ſuſtaining flame; for without it the whole maſs of the atmoſphere would, in time, become unfit for the purpoſe of animal life; and yet there is no reaſon to think that it is, at preſent, at all leſs fit for reſpiration than it it has ever been. I flatter myſelf, however, that I have hit upon two of the methods employed by nature for this great purpoſe. How many others there may be, I cannot tell.

When animals die upon being put into air in which other animals have died, after breathing in it as long as they could, it is plain that the cauſe of their death is not the want of any *pabulum vitæ*, which has been ſuppoſed to be contained in the air, but on account of the air being impregnated with ſomething ſtimulating to their lungs; for they almoſt always die in convulſions, and are ſometimes affected ſo ſuddenly, that they are irrecoverable after a ſingle inſpiration, though they be withdrawn immediately, and every method has been taken to bring them to life again. They are affected in the ſame manner, when they are killed in any other kind of noxious air that I have tried, viz. fixed air, inflammable air, air filled with the fumes of brimſtone, infected with putrid matter, in which a mixture of iron filings and brimſtone has ſtood, or in which charcoal has been burned, or metals calcined, or in nitrous air, &c.

If a mouſe (which is an animal that I have commonly made uſe of for the purpoſe of theſe experiments) can ſtand the firſt ſhock of this ſtimulus, or has been habituated to it by degrees, it will live a conſiderable time in air in which other mice will die inſtantaneouſly. I have frequently found that when a number of mice have been confined in a given quantity of air, leſs than half the time that they have actually lived in it, a freſh mouſe has been inſtantly thrown into convulſions, and died upon being put to them. It is evident, therefore, that if the

the experiment of the Black Hole were to be repeated, a man would ſtand the better chance of ſurviving it, who ſhould enter at the firſt, than at the laſt hour. I have alſo obſerved, that young mice will always live much longer than old ones, or than thoſe which are full grown, when they are confined in the ſame quantity of air. I have ſometimes known a young mouſe to live ſix hours in the ſame circumſtances in which an old mouſe has not lived one. On theſe accounts, experiments with mice, and, for the ſame reaſon, no doubt, with other animals alſo, have a conſiderable degree of uncertainty attending them; and therefore, it is neceſſary to repeat them frequently, before the reſult can be abſolutely depended upon.

The diſcovery of the proviſion in nature for reſtoring air, which has been injured by the reſpiration of animals, having long appeared to me to be one of the moſt important problems in natural philoſophy, I have tried a great variety of ſchemes in order to effect it. In theſe, my guide has generally been to conſider the influences to which the atmoſphere is, in fact, expoſed; and, as ſome of my unſucceſsful trials may be of uſe to thoſe who are diſpoſed to take pains in the farther inveſtigation of this ſubject, I ſhall mention the principal of them.

The noxious effluvium with which air is loaded by animal reſpiration, is not abſorbed by ſtanding without agitation in freſh or ſalt water. I have kept it many months in freſh water, when, inſtead of being meliorated, it has ſeemed to become even more deadly, ſo as to require more time to reſtore it, by the methods which will be explained hereafter, than air which has been lately made noxious. I have even ſpent ſeveral hours in pouring this air from one glaſs veſſel into another, in water, ſometimes as cold, and ſometimes as warm, as my hands could bear it, and have ſometimes alſo wiped the veſſels many times, during the courſe of the experiment, in order to take off that part of the noxious matter, which might adhere to the glaſs veſſels, and which evidently gave them an offenſive ſmell; but all theſe methods were generally without any ſenſible effect. The motion, alſo, which the air received in theſe circumſtances, it is very evident, was of no uſe for this purpoſe.

This kind of air is not reſtored by being expoſed to the light, or by any other influence to which it is expoſed, when confined in a thin phial, in the open air, for ſome months.

Among other experiments, I tried a great variety of different effluvia, which are continually exhaling into the air, eſpecially of thoſe ſubſtances which are known to reſiſt putrefaction; but I could not by theſe means effect any melioration of the noxious quality of this kind of air.

Having read, in the Memoirs of the Imperial Society, of a plague not afflicting a particular village, in which there was a large ſulphur work, I immediately fumigated a quantity of this kind of air; or (which will hereafter appear to be the very ſame thing) air tainted with putrefaction, with the fumes of burning brimſtone, but without any effect.

I once imagined, that the nitrous acid in the air might be the general reſtorative which I was in queſt of; and the conjecture was favoured, by finding

ing that candles would burn, and animals live, in air extracted from ſaltpetre. I therefore ſpent a good deal of time in attempting, by a burning-glaſs, and other means, to impregnate this noxious air with ſome effluvium of ſaltpetre, and, with the ſame view, introduced into it the fumes of the ſmoaking ſpirit of nitre; but both theſe methods were altogether ineffectual.

In order to try the effect of heat, I put a quantity of air, in which mice had died, into a bladder, tied to the end of the ſtem of a tobacco-pipe, at the other end of which was another bladder, out of which the air was carefully preſſed. I then put the middle part of the ſtem into a chafing-diſh of hot coals, ſtrongly urged with a pair of bellows; and, preſſing the bladders alternately, I made the air paſs ſeveral times through the heated part of the pipe. I have alſo made this kind of air very hot, ſtanding in water before the fire. But neither of theſe methods were of any uſe.

Rarefaction and condenſation by inſtruments were alſo tried, but in vain.

Thinking it poſſible that the earth might imbibe the noxious quality of the air, and thence ſupply the roots of plants with ſuch putreſcent matter as is known to be nutritive to them, I kept a quantity of air, in which mice had died, in a phial, one half of which was filled with fine garden mould; but, though it ſtood two months in theſe circumſtances, it was not the better for it.

I once imagined that, ſince ſeveral kinds of air cannot be long ſeparated from common air, by being confined in bladders, in bottles well corked, or even

 cloſed

closed with ground stopples, the affinity between this noxious air and the common air might be so great, that they would mix through a body of water interposed between them; the water continually receiving from the one, and giving to the other, especially as water receives some kinds of impregnation from, I believe, every kind of air to which it is contiguous; but I have seen no reason to conclude, that a mixture of any kind of air with the common air can be produced in this manner. I have kept air in which mice have died, air in which candles have burned out, and inflammable air, separated from the common air, by the slightest partition of water that I could well make, so that it might not evaporate in a day or two, if I should happen not to attend to them; but I found no change in them after a month or six weeks. The inflammable air was still inflammable, mice died instantly in the air in which other mice had died before, and candles would not burn where they had burned out before.

Since air tainted with animal or vegetable putrefaction is the same thing with air rendered noxious by animal respiration, I shall now recite the observations which I have made upon this kind of air, before I treat of the method of restoring them.

That these two kinds of air are, in fact, the same thing, I conclude from their having several remarkable common properties, and from their differing in nothing that I have been able to observe. They equally extinguish flame, they are equally noxious to animals, they are equally, and in the same way, offensive to the smell, they are equally diminished

in

in their quantity, they equally precipitate in lime-water, and they are reſtored by the ſame means.

Since air which has paſſed through the lungs is the ſame thing with air tainted with animal putrefaction, it is probable that one uſe of the lungs is to carry off a putrid effluvium, without which, perhaps, a living body might putrefy as ſoon as a dead one.

When a mouſe putrefies in any given quantity of air, the bulk of it is generally increaſed for a few days; but in a few days more it begins to ſhrink up, and generally, in about eight or ten days, if the weather be pretty warm, it will be found to be diminiſhed $\frac{1}{6}$, or $\frac{1}{5}$ of its bulk. If it do not appear to be diminiſhed after this time, it only requires to be paſſed through water, and the diminution will not fail to be ſenſible. I have ſometimes known almoſt the whole diminution to take place, upon once or twice paſſing through the water. The ſame is the caſe with air, in which animals have breathed as long as they could. Alſo, air in which candles have burned out may almoſt always be farther reduced by this means. All theſe proceſſes, as I obſerved before, ſeem to diſpoſe the compound maſs of air to part with ſome conſtituent part belonging to it; and this being miſcible with water, muſt be brought into contact with it, in order to mix with it to the moſt advantage, eſpecially when its union with the other conſtituent principles of the air is but partially broken.

I have put mice into veſſels which had their mouths immerſed in quickſilver, and obſerved that the air was not much contracted after they were dead or cold; but upon withdrawing the mice, and admitting lime-water to the air it immediately became turbid, and was contracted in its dimensions as usual.

I tried the same thing with air tainted with putrefaction, putting a dead mouse to a quantity of common air, in a vessel which had its mouth immersed in quicksilver, and after a week I took the mouse out, drawing it through the quicksilver, and observed that for some time there was an apparent increase of the air perhaps about $\frac{1}{30}$. After this, it stood two days in the quicksilver, without any sensible alteration; and then admitting water to it, it began to be absorbed, and continued so, till the original quantity was diminished about $\frac{1}{6}$. If, instead of common water, I had made use of lime-water in this experiment, I make no doubt but it would have become turbid.

If a quantity of lime-water in a phial be put under a glass vessel standing in water, it will not become turbid, and provided the access of the common air be prevented, it will continue lime-water, I do not know how long; but if a mouse be left to putrefy in the vessel, the water will deposit all its lime in a few days. This may be owing to the fixed air being transferred from the putrid mouse into the water, and yet it is evident that there is a putrid effluvium intirely distinct from this kind of air, and which has very different properties.

It is a doubt with me, however, whether the putrid effluvium be not chiefly fixed air, with the addition of some other effluvium, which has the power of diminishing common air. The resemblance between the true putrid effluvium and fixed air in the following experiment, which is as decisive

as

as I can poſſibly contrive it, appeared to be very great; indeed, much greater than I had expected. I put a dead mouſe into a tall glaſs veſſel, and having filled the remainder with quickſilver, and ſet it, inverted, in a pot of quickſilver, I let it ſtand about two months, in which time the putrid effluvium iſſuing from the mouſe had filled the whole veſſel, and part of the diſſolved blood, which lodged upon the ſurface of the quikſilver, began to be thrown out. I then filled another glaſs veſſel, of the ſame ſize and ſhape, with as pure fixed air as I could make, and expoſed them both, at the ſame time, to a quantity of lime-water. In both caſes the water grew turbid alike, it roſe equally faſt in both the veſſels, and likewiſe equally high; ſo that about the ſame quantity remained unabſorbed by the water. One of theſe kinds of air, however, was exceedingly ſweet and pleaſant, and the other inſufferably offenſive; one of them alſo would have made an addition to any quantity of common air with which it had been mixed, and the other would have diminiſhed it. This, at leaſt, would have been the conſequence, if the mouſe itſelf had putrefied in any quantity of air.

It ſeems to depend, in ſome meaſure, upon the time, and other circumſtances, in the diſſolution of animal or vegetable ſubſtances, whether they yield the proper putrid effluvium, or fixed, or inflammable air; but the experiments which I have made upon this ſubject, have not been numerous enough to enable me to decide with certainty concerning thoſe circumſtances. Putrid cabbage, green, or boiled, infects the air in the very ſame manner as putrid animal ſubſtances. Air thus tainted is equally contracted

in

in its dimensions, it equally extinguishes flame, and is equally noxious to animals; but they affect the air very differently if the heat that is applied to them be considerable. If beef or mutton, raw, or boiled, be placed so near to the fire, that the heat to which it is exposed shall equal, or rather exceed, that of the blood, a considerable quantity of air will be generated in a day or two, about $\frac{1}{7}$th of which I have generally found to be absorbed by water, while all the rest was inflammable; but air generated from vegetables, in the same circumstances, will be almost all fixed, and no part of it inflammable. This I have repeated again and again, the whole process being in quicksilver; so that neither common air, nor water, had any access to the substance on which the experiment was made; and the generation of air, or effluvium of any kind, except what might be absorbed by quicksilver, or resorbed by the substance itself, might be distinctly noted.

A vegetable substance, after standing a day or two in these circumstances, will yield nearly all the air that can be extracted from it, in that degree of heat; whereas an animal substance will continue to give more air or effluvium, of some kind or other, with very little alteration, for many weeks. It is remarkable, however, that though a piece of beef or mutton, plunged in quicksilver, and kept in this degree of heat, yield air, the bulk of which is inflammable, and contracts no putrid smell (at least, in a day or two), a mouse treated in the same manner, yields the proper putrid effluvium, as, indeed the smell sufficiently indicates; and this effluvium does

either

either itſelf extinguiſh flame, or has in it ſuch a mixture of fixed air, as to give it that property.

That the putrid effluvium will mix with water ſeems to be evident from the following experiment. If a mouſe be put into a jar full of water, ſtanding with its mouth inverted in another veſſel of water, a conſiderable quantity of elaſtic matter (and which may, therefore, be called air) will ſoon be generated, unleſs the weather be ſo cold as to check all putrefaction. After a ſhort time, the water contracts an extremely fetid and offenſive ſmell, which ſeems to indicate that the putrid effluvium pervades the water, and affects the neighbouring air; and ſince, after this, there is often no increaſe of the air, that ſeems to be the very ſubſtance which is carried off through the water, as faſt as it is generated; and the offenſive ſmell is a ſufficient proof that it is not fixed air. For this has a very agreeable flavour, whether it be produced by fermentation, or extracted from chalk by oil of vitriol; affecting not only the mouth, but even the noſtrils, with a pungency which is peculiarly pleaſing to a certain degree, as any perſon may eaſily ſatisfy himſelf who will chuſe to make the experiment. If the water in which the mouſe was immerſed, and which is ſaturated with the putrid air, be changed, the greater part of the putrid air will, in a day or two, be abſorbed, though the mouſe continues to yield the putrid effluvium as before; for as ſoon as this freſh water becomes ſaturated with it, it begins to be offenſive to the ſmell, and the quantity of the putrid air upon its ſurface increaſes as before. I kept a mouſe producing putrid air in this manner for the ſpace of ſeveral months.

Six

Six ounce meaſures of air not readily abſorbed by water, appeared to have been generated from one mouſe, which had been putrefying eleven days in confined air, before it was put into a jar which was quite filled with water, for the purpoſe of this obſervation.

Air thus generated from putrid mice ſtanding in water, without any mixture of common air, extinguiſhes flame, and is noxious to animals, but not more ſo than common air only tainted with putrefaction. It is exceedingly difficult and tedious to collect a quantity of this putrid air, not miſcible in water, ſo very great a proportion of what is collected being abſorbed by the water, in which it is kept; but what that proportion is, I have not endeavoured to aſcertain.

Though a quantity of air be diminiſhed by any ſubſtance putrefying in it, I have not yet found the ſame effect to be produced by a mixture of putrid air with common air; but, in the manner in which I have hitherto made the experiment, I was obliged to let the putrid air, paſs through a body of water; which might inſtantly abſorb whatever it was in the putrid ſubſtance, that diminiſhed the common air.

Inſects of various kinds live perfectly well in air tainted with animal or vegetable putrefaction, when a ſingle inſpiration of it would have inſtantly killed any animal. I have frequently tried the experiment with flies and butterflies. I have alſo obſerved, that the *aphides* will thrive as well upon plants growing in this kind of air, as in the open air. I have even been frequently obliged to take plants out of the putrid air in which they were growing, on purpoſe to bruſh away the ſwarms of thefe infects which infected them; and yet fo effectually did fome of them conceal themfelves, and fo faft did they multiply, in thefe circumftances, that I could feldom keep the plants quite clear of them.

When air has been frefhly and ftrongly tainted with putrefaction, fo as to fmell through the water, fprigs of mint have prefently died, upon being put into it, their leaves turning black; but if they do not die prefently, they thrive in a moft furprizing manner. In no other circumftances have I ever feen vegetation fo vigorous as in this kind of air, which is immediately fatal to animal life. Though thefe plants have been crouded in jars filled with this air, every leaf has been full of life; frefh fhoots have branched out in various directions, and have grown much fafter than other fimilar plants, growing in the fame expofure in common air.

This obfervation led me to conclude, that plants, inftead of affecting the air in the fame manner with animal refpiration, reverfe the effects of breathing, and tend to keep the atmofphere fweet and wholefome, when it is become noxious, in confequence of animals living and breathing, or dying and putrefying in it.

In order to afcertain this, I took a quantity of air, made thoroughly noxious, by mice breathing and dying in it, and divided it into two parts; one of which I put into a phial immerfed in water; and to the other (which was contained in a glafs jar, ftanding in water) I put a fprig of mint. This was about the beginning of Auguft 1771, and after eight or nine days, I found that a moufe lived perfectly well in in that part of the air, in which the ſprig of mint had grown, but died the moment it was put into the other part of the ſame original quantity of air; and which I had kept in the very ſame expoſure, but without any plant growing in it.

This experiment I have ſeveral times repeated; ſometimes uſing air, in which animals had breathed and died; ſometimes uſing air tainted with vegetable or animal putrefaction, and generally with the ſame ſucceſs.

Once, I let a mouſe live and die in a quantity of air, which had been noxious, but which had been reſtored by this proceſs, and it lived nearly as long as I conjectured it might have done in an equal quantity of freſh air; but, this is ſo exceedingly various, that it is not eaſy to form any judgment from it; and in this caſe the ſymptom of *difficult reſpiration* ſeemed to begin earlier than it would have done in common air.

Since the plants that I made uſe of manifeſtly grow and thrive in putrid air; ſince putrid matter is well known to afford proper nouriſhment for the roots of plants; and ſince it is likewiſe certain that they receive nouriſhment by their leaves as well as by their roots, it ſeems to be exceedingly probable, that the putrid effluvium is in ſome meaſure extracted from the air, by means of the leaves of plants, and therefore that they render the remainder more fit for reſpiration.

Towards the end of the year ſome experiments of this kind did not anſwer ſo well as they had done before, and I had inſtances of the relapſing of this reſtored air to its former noxious ſtate. I therefore

ſuſpended

ſuſpended my judgment concerning the efficacy of plants to reſtore this kind of noxious air, till I ſhould have an opportunity of repeating my experiments, and giving more attention to them. Accordingly I reſumed the experiments in the ſummer of the year 1772, when I preſently had the moſt indiſputable proof of the reſtoration of putrid air by vegetation; and as the fact is of ſome importance, and the ſubſequent variation in the ſtate of this kind of air is a little remarkable; I think it neceſſary to relate ſome of the facts pretty circumſtantially.

The air, on which I made the firſt experiments, was rendered exceedingly noxious by mice dying in it on the 20th of June. Into a jar nearly filled with one part of this air, I put a ſprig of mint, while I kept another part of it in a phial, in the ſame expoſure; and on the 27th of the ſame month, and not before, I made a trial of it, by introducing a mouſe into a glaſs veſſel, containing 2½ ounce meaſures filled with each kind of air; and I noted the following facts.

When the veſſel was filled with the air in which the mint had grown, a very large mouſe lived five minutes in it, before it began to ſhew any ſign of uneaſineſs. I then took it out, and found it to be as ſtrong and vigorous as when it was firſt put in; whereas in that air which had been kept in the phial only, without a plant growing in it, a younger mouſe continued not longer than two or three ſeconds, and was taken out quite dead. It never breathed after, and was immediately motionleſs. After half an hour, in which time the larger mouſe

 (which

(which I had kept alive, that the experiment might be made on both the kinds of air with the very ſame animal) would have been ſufficiently recruited, ſuppoſing it to have received any injury by the former experiment, was put into the ſame veſſel of air; but though it was withdrawn again, after being in it hardly one ſecond, it was recovered with difficulty, not being able to ſtir from the place for near a minute. After two days, I put the ſame mouſe into an equal quantity of common air, and obſerved that it continued ſeven minutes without any ſign of uneaſineſs; and being very uneaſy after three minutes longer, I took it out. Upon the whole, I concluded that the reſtored air wanted about one fourth of being as wholeſome as common air. The ſame thing alſo appeared when I applied the teſt of nitrous air.

In the ſeven days, in which the mint was growing in this jar of noxious air, three old ſhoots had extended themſelves about three inches, and ſeveral new ones had made their appearance in the ſame time. Dr. Franklin and Sir John Pringle happened to be with me, when the plant had been three or four days in this ſtate, and took notice of its vigorous vegetation, and remarkably healthy appearance in that confinement.

On the 30th of the ſame month, a mouſe lived fourteen minutes, breathing naturally all the time, and without appearing to be much uneaſy, till the laſt two minutes, in air which had been rendered noxious by mice breathing in it almoſt a year before, and which I had found to be moſt highly noxious on the 19th of this month, a plant having grown in it, but

but not exceedingly well, theſe eleven days; on which account, I had deferred making the trial ſo long. This reſtored air was affected by a mixture of nitrous air, almoſt as much as common air.

As this putrid air was thus eaſily reſtored to a conſiderable degree of fitneſs for reſpiration, by plants growing in it, I was in hopes that by the ſame means it might in time be ſo much more perfectly reſtored, that a candle would burn in it; and for this purpoſe I kept plants growing in the jars which contained this air till the middle of Auguſt following, but did not take ſufficient care to pull out all the old and rotten leaves. The plants, however, had grown, and looked ſo well upon the whole, that I had no doubt but that the air muſt conſtantly have been in a mending ſtate; when I was exceedingly ſurprized to find, on the 24th of that month, that though the air in one of the jars had not grown worſe, it was no better, and that the air in the other jar was ſo much worſe than it had been, that a mouſe would have died in it in a few ſeconds. It alſo made no effervеſcence with nitrous air, as it had done before.

Suſpecting that the ſame plant might be capable of reſtoring putrid air to a certain degree only, or that plants might have a contrary tendency in ſome ſtages of their growth, I withdrew the old plant, and put a freſh one in its place; and found that, after ſeven days, the air was reſtored to its former wholeſome ſtate. This fact I conſider as a very remarkable one, and well deſerving of a farther inveſtigation, as it may throw more light upon the principles of vegetation. It is not, however, a ſingle fact; for I had ſeveral inſtances of the ſame kind in the preceding year; but it ſeemed ſo very extraordinary, that air ſhould grow worſe by the continuance of the ſame treatment by which it had grown better, that, whenever I obſerved it, I concluded that I had not taken ſufficient care to ſatisfy myſelf of its previous reſtoration.

That plants are capable of perfectly reſtoring air injured by reſpiration, may, I think, be inferred with certainty from the perfect reſtoration, by this means, of air which had paſſed through my lungs, ſo that a candle would burn in it again, though it had extinguiſhed flame before, and a part of the ſame original quantity of air ſtill continued to do ſo. Of this one inſtance occurred in the year 1771, a ſprig of mint having grown in a jar of this kind of air, from the 25th of July to the 17th of Auguſt following; and another trial I made with the ſame ſucceſs the 7th of July 1772, the plant having grown in it from the 29th of June preceding. In this caſe alſo I found that the effect was not owing to any virtue in the leaves of mint; for I kept them conſtantly changed in a quantity of this kind of air, for a conſiderable time, without making any ſenſible alteration in it.

Theſe proofs of a partial reſtoration of air by plants in a ſtate of vegetation, though in a confined and unnatural ſituation, cannot but render it highly probable, that the injury which is continually done to the atmoſphere by the reſpiration of ſuch a number of animals, and the putrefaction of ſuch maſſes of both vegetable and animal matter, is, in part at leaſt, repaired by the vegetable creation.

And,

And, notwithſtanding the prodigious maſs of air that is corrupted daily by the abovementioned cauſes; yet, if we conſider the immenſe profuſion of vegetables upon the face of the earth, growing in places ſuited to their nature, and conſequently at full liberty to exert all their powers, both inhaling and exhaling, it can hardly be thought, but that it may be a ſufficient counterbalance to it, and that the remedy is adequate to the evil.

Dr. Franklin, who, as I have already obſerved, ſaw ſome of my plants in a very flouriſhing ſtate, in highly noxious air, was pleaſed to expreſs very great ſatisfaction with the reſult of the experiments. In his anſwer to the letter in which I informed him of it, he ſays,

"That the vegetable creation ſhould reſtore the "air which is ſpoiled by the animal part of it, "looks like a rational ſyſtem, and ſeems to be of "a piece with the reſt. Thus fire purifies water "all the world over. It purifies it by diſtillation, "when it raiſes it in vapours, and lets it fall in "rain; and farther ſtill by filtration, when, keep-"ing it fluid, it ſuffers that rain to percolate the "earth. We knew before, that putrid animal ſub-"ſtances were converted into ſweet vegetables, "when mixed with the earth, and applied as "manure; and now, it ſeems, that the ſame pu-"trid ſubſtances, mixed with the air, have a ſimi-"lar effect. The ſtrong thriving ſtate of your "mint in putrid air ſeems to ſhew that the air is "mended by taking ſomething from it, and not "by adding to it." He adds, "I hope this will "give ſome check to the rage of deſtroying trees

" that grow near houſes, which has accompanied " our late improvements in gardening, from an " opinion of their being unwholeſome. I am certain, from long obſervation, that there is nothing unhealthy in the air of woods; for we " Americans have every where our country habitations in the midſt of woods, and no people on " earth enjoy better health, or are more prolific."

Having rendered inflammable air perfectly innoxious by continued agitation in a trough of water, deprived of its air, I concluded that other kinds of noxious air might be reſtored by the ſame means; and I preſently found that this was the caſe with putrid air, even of more than a year's ſtanding. I ſhall obſerve once for all, that this proceſs has never failed to reſtore any kind of noxious air on which I have tried it, viz. air injured by reſpiration or putrefaction, air infected with the fumes of burning charcoal, and of calcined metals, air in which a mixture of iron filings and brimſtone, or that in which paint made of white lead and oil has ſtood, or air which has been diminiſhed by a mixture of nitrous air. Of the remarkable effect which this proceſs has on nitrous air itſelf, an account will be given in its proper place.

If this proceſs be made in water deprived of air, either by the air pump, by boiling, by diſtillation, or if freſh rain water be uſed, the air will always be diminiſhed by the agitation; and this is certainly the faireſt method of making the experiment. If the water be freſh pump water, there will always be an increaſe of the air by agitation, the air contained in the water being ſet looſe, and joining

joining that which is in the jar. In this caſe, alſo, the air has never failed to be reſtored; but then it might be ſuſpected that the melioration was produced by the addition of ſome more wholeſome ingredient. As theſe agitations were made in jars with wide mouths, and in a trough which had a large ſurface expoſed to the common air, I take it for granted that the noxious effluvia, whatever they be, were firſt imbibed by the water, and thereby tranſmitted to the common atmoſphere. In ſome caſes this was ſufficiently indicated by the diſagreeable ſmell which attended the operation.

After I had made theſe experiments, I was informed that an ingenious phyſician and philoſopher had kept a fowl alive twenty-four hour, in a quantity of air in which another fowl of the ſame ſize had not been able to live longer than an hour, by contriving to make the air, which it breathed, paſs through no very large quantity of acidulated water, the ſurface of which was not expoſed to the common air; and that even when the water was not acidulated, the fowl lived much longer than it could have done, if the air which it breathed had not been drawn through the water. As I ſhould not have concluded that this experiment would have ſucceeded ſo well, from any obſervations that I had made upon the ſubject, I took a quantity of air in which mice had died, and agitated it very ſtrongly, firſt in about five times its own quantity of diſtilled water, in the manner in which I had impregnated water with fixed air; but though the operation was continued a long time, it made no ſenſible change in the properties of the air. I alſo repeated the operation with

 pump

pump water, but with as little effect. In this caſe, however, though the air was agitated in a phial, which had a narrow neck, the ſurface of the water in the baſon was conſiderably large, and expoſed to the common atmoſphere, which muſt have tended a little to favour the experiment. In order to judge more preciſely of the effect of theſe different methods of agitating air, I transfered the very noxious air, which I had not been able to amend in the leaſt degree by the former method, into an open jar, ſtanding in a trough of water; and when I had agitated it till it was diminiſhed about one third, I found it to be better than air, in which candles had burned out, as appeared by the teſt of the nitrous air; and a mouſe lived in 2½ ounce meaſures of it a quarter of an hour, and was not ſenſibly affected the firſt ten or twelve minutes.

In order to determine whether the addition of any *acid* to the water, would make it more capable of reſtoring putrid air, I agitated a quantity of it in a phial containing very ſtrong vinegar; and after that in *aqua fortis*, only half diluted with water; but, by neither of theſe proceſſes was the air at all mended, though the agitation was repeated at intervals during a whole day, and it was moreover allowed to ſtand in that ſituation all night.

Since, however, water in theſe experiments muſt have imbibed and retained a certain portion of the noxious effluvia, before they could be tranſmited to the external air, I do not think it improbable but that the agitation of the ſea and large lakes may be of ſome uſe for the purification of the atmoſphere, and the putrid matter contained in water may be

imbibed

imbibed by aquatic plants, or be deposited in some other manner.

Having found, by several experiments abovementioned, that the proper putrid effluvium is something quite distinct from fixed air, and finding, by the experiments of Dr. Macbride, that fixed air corrects putrefaction; I once concluded that this effect was produced, not by stopping the flight of the fixed air, or restoring to the putrefying substance the very same thing that had escaped from it; and which was the common vinculum of all its parts (which is that ingenious author's hypothesis) but by an affinity between the fixed air and the putrid effluvium. It therefore occurred to me, that fixed air, and air tainted with putrefaction, though equally noxious when separate, might make a wholesome mixture, the one correcting the other; and I was confirmed in this opinion by, I believe, not less than fifty or sixty instances, in which air, that had been made in the highest degree noxious, by respiration or putrefaction, was so far sweetened, by a mixture of about four times as much fixed air that afterwards mice lived in it exceedingly well, and in some cases almost as long as in common air. I found it, indeed, to be more difficult to restore old putrid air by this means; but I hardly ever failed to do it, when the two kinds of air had stood a long time together, by which I mean about a fortnight or three weeks.

The reason why I do not absolutely conclude that the restoration of air in these cases was the effect of fixed air, is that, when I made a trial of the mixture, I sometimes agitated the two kinds of air pretty ſtrongly together, in a trough of water, or at leaſt paſſed it ſeveral times through the water, from one jar to another, that the ſuperfluous fixed air might be abſorbed, not ſuſpecting at that time that the agitation could have any other effect; but having ſince found that very violent, and eſpecially long continued agitation in water, without any mixture of fixed air, never failed to render any kind of noxious air in ſome meaſure fit for reſpiration (and in one particular inſtance the mere transferring of the air from one veſſel to another through the water, though for a much longer time than I ever uſed for the mixtures of air, was of conſiderable uſe for the ſame purpoſe); I began to entertain ſome doubt of the efficacy of fixed air, for that purpoſe. In ſome caſes alſo the mixture of fixed air had by no means ſo much effect on the putrid air as, from the generality of my obſervations, I ſhould have expected.

I was always aware, indeed, that it might be ſaid, that, the reſiduum of fixed air not being very noxious, ſuch an addition muſt contribute to mend the putrid air; but, in order to obviate this objection, I once mixed the reſiduum of as much fixed air as I had found, by a variety of trials, to be ſufficient to reſtore a given quantity of putrid air, with an equal quantity of putrid air, without making any ſenſible melioration of it.

Upon the whole, I am inclined to think that this proceſs could hardly have ſucceeded ſo well as it did with me, and in ſo great a number of trials, unleſs fixed air have ſome tendency to correct air tainted with reſpiration or putrefaction; and it is perfectly agreeable to the analogy of Dr. Macbride's discoveries, and may naturally be expected from them, that it should have such an effect.

By a mixture of fixed air I have made wholesome the residuum of air generated by putrefaction only, from mice plunged in water. This, one would imagine, *à priori*, to be the most noxious of all kinds of air. For if common air only tainted with putrefaction be so deadly, much more might one expect that air to be so, which was generated from putrefaction only; but it seems to be nothing more than common air tainted with putrefaction, and therefore requires no other process to sweeten it. In this case, however, we seem to have an instance of the generation of genuine common air, though mixed with something that is foreign to it. Perhaps the residuum of fixed air may be another instance of the same nature.

Fixed air is equally diffused through the whole mass of any quantity of putrid air with which it is mixed; for dividing the mixture into two equal parts, they were reduced in the same proportion by passing through water. But this is also the case with some of the kinds of air which will not incorporate, as inflammable air, and air in which brimstone has burned.

If fixed air tend to correct air which has been injured by animal respiration or putrefaction, limekilns, which discharge great quantities of fixed air, may be wholesome in the neighbourhood of populous cities, the atmosphere of which must abound with putrid effluvia. I should think also that physicians might avail themselves of the application

of

of fixed air in many putrid diſorders, eſpecially as it may be ſo eaſily adminiſtered by way of clyſter, where it would often find its way to much of the putrid matter. Nothing is to be apprehended from the diſtention of the bowels by this kind of air, ſince it is ſo readily abſorbed by any fluid or moiſt ſubſtance. Since fixed air is not noxious *per ſe*, but, like fire, only in exceſs, I do not think it at all hazardous to attempt to breathe it. It is however eaſily conveyed into the ſtomach, in natural or artificial Pyrmont water, in briſkly fermenting liquors, or a vegetable diet. It is poſſible, however, that a conſiderable quantity of fixed air might be imbibed by the abſorbing veſſels of the ſkin, if the whole body, except the head, ſhould be ſuſpended over a veſſel of ſtrongly fermenting liquor; and in ſome putrid diſorders this treatment might be very ſalutary. If the body was expoſed quite naked, there would be very little danger from the cold in this ſituation, and the air having freer acceſs to the ſkin might produce a greater effect. Being no phyſician, I run no riſk by throwing out theſe random, and perhaps whimſical, propoſals.

Having communicated my obſervations on fixed air, and eſpecially my ſcheme of applying it by way of *clyſter* in putrid diſorders, to Mr. Hey, an ingenious ſurgeon in this town, a caſe preſently occurred, in which he had an opportunity of giving it a trial; and mentioning it to Dr. Hird and Dr. Crowther, two phyſicians who attended the patient, they approved the ſcheme, and it was put in execution: both by applying the fixed air by way of clyſter, and at the ſame time making the patient drink plentifully of liquors ſtrongly impregnated with it. The event was ſuch, that I requeſted Mr. Hey to draw up a particular account of the caſe, deſcribing the whole of the treatment, that the public might be ſatisfied that this new application of fixed air is perfectly ſafe, and alſo have an opportunity of judging how far it had the effect which I expected from it; and as the application is new, and not unpromiſing, I ſhall beg leave to ſubjoin his letter to me on the ſubject, by way of *Appendix* to theſe papers.

V.

Of Air in which a mixture of brimstone and filings of iron has stood.

Finding in Dr. Hales's account of his experiments, that there was a great diminution of the quantity of air in which a mixture of powdered brimſtone and filings of iron, made into a paſte with water, had ſtood, I repeated the experiment; and found the diminution greater than I had expected. The diminution of air by this proceſs is made as effectually, and as expeditiouſly, in quickſilver as in water; and it may be meaſured with the greateſt accuracy, becauſe there is neither any previous expanſion nor increaſe of the quantity of air, and becauſe it is ſome time before it begins to have any ſenſible effect. The diminution of air by this proceſs is various; but I have generally

generally found it to be between $\frac{1}{4}$ and $\frac{1}{5}$ of the whole.

Air thus diminiſhed is not heavier, but rather lighter than common air; and though lime-water does not become turbid when it is expoſed to this air, it is probably owing to the formation of a ſelenitic ſalt, as was the caſe with the ſimple burning of brimſtone abovementioned. That ſomething proceeding from the brimſtone ſtrongly affects the water which is confined in the ſame place with this brimſtone, is manifeſt from the very ſtrong ſmell that it has of the volatile ſpirit of vitriol. I conclude the diminution of air by this proceſs is of the ſame kind with the diminution of it in the other caſes, becauſe when this mixture is put into air which has been previouſly diminiſhed, either by the burning of candles, by reſpiration, or putrefaction, though it never fails to diminiſh it ſomething more, it is, however, no farther than this proceſs alone would have done it. If a freſh mixture be introduced into a quantity of air which had been reduced by a former mixture, it has little or no farther effect.

I obſerved, that when a mixture of this kind was taken out of a quantity of air in which a candle had before burned out, and in which it had ſtood for ſeveral days, it was quite cold and black, as it always becomes in a confined place; but it preſently grew very hot, ſmoaked copiouſly, and ſmelled very offenſively; and when it was cold, it was brown, like the ruſt of iron.

I once put a mixture of this kind to a quantity of inflammable air, made from iron, by which means it was diminiſhed $\frac{1}{9}$ or $\frac{1}{10}$ in its bulk; but, as far as I could

I could judge, it was ſtill as inflammable as ever. Another quantity of inflammable air was alſo reduced in the ſame proportion, by a mouſe putrefying in it; but its inflammability was not ſeemingly leſſened.

Air diminiſhed by this mixture of iron filings and brimſtone, is exceedingly noxious to animals, and I have not perceived that it grows any better by keeping in water. The ſmell of it is very pungent and offenſive.

The quantity of this mixture which I made uſe of in the preceding experiments, was from two to four ounce meaſures; but I did not perceive, but that the diminution of the quantity of air (which was generally about twenty ounce meaſures) was as great with the ſmalleſt, as with the largeſt quantity. How ſmall a quantity is neceſſary to diminiſh a given quantity of air to a *maximum*, I have made no experiments to aſcertain.

As ſoon as this mixture of iron filings, with brimſtone and water, begins to ferment, it alſo turns black, and begins to ſwell, and it continues to do ſo, till it occupies twice as much ſpace as it did at firſt; and the force with which it expands is great; but how great it is I have not endeavoured to determine.

When this mixture is immerſed in water, it generates no air, though it becomes black, and ſwells.

VI.

Of nitrous Air.

Ever ſince I firſt read Dr. Hales's moſt excellent Statical Eſſays, I was particularly ſtruck with that experiment of his, of which an account is given, Vol. I. p. 224, and Vol. II. p. 280; in which common air, and air generated from the Walton pyrites, by ſpirit of nitre, made a turbid red mixture, and in which part of the common air was abſorbed; but I never expected to have the ſatisfaction of ſeeing this remarkable appearance, ſuppoſing it to be peculiar to that particular mineral. Happening to mention this ſubject to the Hon. Mr. Cavendiſh, when I was in London, in the ſpring of the year 1772, he ſaid that he did not imagine but that other kinds of pyrites might anſwer as well as that which Dr. Hales made uſe of, and that probably the red appearance of the mixture depended upon the ſpirit of nitre only. This encouraged me to attend to the ſubject; and having no pyrites, I began with the ſolution of the different metals in ſpirit of nitre, and catching the air which was generated in the ſolution, I preſently found what I wanted, and a good deal more.

Beginning with the ſolution of braſs, on the 4th of June 1772, I firſt found this remarkable ſpecies of air; one effect of which, though it was caſually obſerved by Dr. Hales, he gave but little attention to; and which, as far as I know, has paſſed altogether unnoticed ſince his time, inſomuch that no name has been given to it. I therefore found myſelf, contrary

to

to my firſt reſolution, under an abſolute neceſſity of giving a name to this kind of air myſelf. When I firſt began to ſpeak and write of it to my friends, I happened to diſtinguiſh it by the name of nitrous air, becauſe I had procured it by means of ſpirit of nitre only; and though I cannot ſay that I altogether like the term, becauſe this air is not got from all the metals by the ſame ſpirit, neither myſelf nor any of my friends, to whom I have applied for the purpoſe, have been able to hit upon a better; ſo that I am obliged, after all, to content myſelf with it.

I have found that this kind of air is readily procured from iron, copper, braſs, tin, ſilver, quickſilver, biſmuth, and nickel, by the nitrous acid only, and from gold and the regulus of antimony by aqua regia. The circumſtances attending the ſolution of each of theſe metals are various, but hardly worth mentioning, in treating of the properties of the air which they yield, which, from what metal ſoever it is extracted, has, as far as I have been able to obſerve, the very ſame properties.

One of the moſt conſpicuous properties of this kind of air is the great diminution of any quantity of common air with which it is mixed, attended with a turbid red, or deep orange colour, and a conſiderable heat. The ſmell of it, alſo, is very ſtrong, and remarkable, but very much reſembling that of ſmoking ſpirit of nitre.

The diminution of a mixture of this and common air is not an equal diminution of both the kinds, which is all that Dr. Hales could obſerve, but of the common air chiefly, though not wholly. For if one meaſure of nitrous air be put to two meaſures of common air, in a few minutes (by which time the effervescence will be over, and the mixture will have recovered its transparency) there will want about one ninth of the original two measures. I hardly know any experiment that is more adapted to amaze and surprize than this is, which exhibits a quantity of air, which, as it were, devours a quantity of another kind of air half as large as itself, and yet is so far from gaining any addition to its bulk, that it is diminished by it. If, after this full saturation of common air with nitrous air, more nitrous air be put to it, it makes an addition equal to its own bulk, without producing the least redness, or any other visible effect.

That this diminution is chiefly in the quantity of common air, is evident from this observation, that if the smallest quantity of common air be put to any larger quantity of nitrous air, though the two together will not occupy so much space as they did separately, yet the quantity will be still larger than that of the nitrous air only. One ounce measure of common air being put to near twenty ounce measures of nitrous air, made an addition to it of about half an ounce measure. This, however, being a much greater proportion than the diminution of common air, in the former experiment, seems to prove that part of the diminution in the former case is in the nitrous air. Besides, it will presently appear, that nitrous air is subject to a most remarkable diminution; and as common air, in a variety of other cases, suffers a diminution from one fifth to one fourth, I conclude, that in this case also it does not exceed that proportion, and therefore that the remainder of the diminution respects the nitrous air.

In

In order to judge whether the water contributed to the diminution of this mixture of nitrous and common air, I made the whole proceſs ſeveral times in quickſilver, uſing one third of nitrous, and two thirds of common air, as before. In this caſe the redneſs continued a very long time, and the diminution was not ſo great as when the mixtures had been made in water, there remaining one ſeventh more than the original quantity of common air. This mixture ſtood all night upon the quickſilver; and the next morning I obſerved that it was no farther diminiſhed upon the admiſſion of water to it, nor by pouring it ſeveral times through the water, and letting it ſtand in water two days. Another mixture, which ſtood about ſix hours on the quickſilver, was diminiſhed a little more upon the admiſſion of water, but was never leſs than the original quantity of common air. In another caſe, however, in which the mixture ſtood but a very ſhort time in quickſilver, the farther diminution, which took place upon the admiſſion of water, was much more conſiderable; ſo that the diminution, upon the whole, was very nearly as great as if the proceſs had been intirely in water. It is evident from theſe experiments, that the diminution is in part owing to the abſorption by the water; but that when the mixture is kept a long time, in a ſituation in which there is no water to abſorb any part of it, it acquires a conſtitution, by which it is afterwards incapable of being abſorbed by water.

In order to determine whether the fixed part of common air was depoſited in the diminution of it by by nitrous air, I inclosed a vessel full of lime water in the jar in which the process was made, but it occasioned no precipitation of the lime; and when the vessel was taken out, after it had been in that situation a whole day, the lime was easily precipitated by breathing into it as usual.

It is exceedingly remarkable that this effervescence and diminution, occasioned by the mixture of nitrous air, is peculiar to common air, or air fit for respiration; and, as far as I can judge, from a great number of observations, is at least very nearly, if not exactly, in proportion to its fitness for this purpose; so that by this means the goodness of air may be distinguished much more accurately than it can be done by putting mice, or any other animals, to breathe in it. This was a most agreeable discovery to me, as I hope it may be an useful one to the public; especially as, from this time, I had no occasion for so large a stock of mice as I had been used to keep for the purpose of these experiments, using them only in those which required to be very decisive; and in these cases I have seldom failed to know beforehand in what manner they would be affected.

It is also remarkable that, on whatever account air is unfit for respiration, this same test is equally applicable. Thus there is not the least effervescence between nitrous and fixed air, or inflammable air, or any species of diminished air. Also the degree of diminution being from nothing at all to more than one third of the whole of any quantity of air, we are by this means in possession of a prodigiously large scale, by which we may distinguish very ſmall degrees of difference in the goodneſs of air. I have not attended much to this circumſtance, having uſed this teſt chiefly for greater differences; but, if I did not deceive myſelf, I have perceived a real difference in the air of my ſtudy, after a few perſons have been with me in it, and the air on the outſide of the houſe. Alſo a phial of air having been ſent me, from the neighbourhood of York, it appeared not to be ſo good as the air near Leeds; that is, it was not diminiſhed ſo much by an equal mixture of nitrous air, every other circumſtance being as nearly the ſame as I could contrive. It may perhaps be poſſible, but I have not yet attempted it, to diſtinguiſh ſome of the different winds, or the air of different times of the year, by this teſt.

By means of this teſt I was able to determine what I was before in doubt about, *viz.* the kind as well as the degree of injury done to air by candles burning in it. I could not tell with certainty by means of mice, whether it was at all injured with reſpect to reſpiration; and yet if nitrous air may be depended upon for furniſhing an accurate teſt, it muſt be rather more than one third worſe than common air, and have been diminiſhed by the ſame general cauſe of the other diminutions of air. For when, after many trials, I put one meaſure of thoroughly putrid and highly noxious air, into the ſame veſſel with two meaſures of good wholeſome air, and into another veſſel an equal quantity, *viz.* three meaſures of air in which a candle had burned out; and then put equal quantities of nitrous air to each of them, the former was diminiſhed rather more than the latter. It agrees with

with this obſervation, that burned air is farther dimininiſhed both by putrefaction, and a mixture of iron filings and brimſtone; and I therefore, take it for granted, by every other cauſe of the diminution of air. It is probable, therefore, that burned air is air ſo far loaded with phlogiſton, as to be able to extinguiſh a candle, which it may do long before it is fully ſaturated.

Inflammable air with a mixture of nitrous air burns with a green flame. This makes a very pleaſing experiment when it is properly conducted. As, for ſome time, I chiefly made uſe of copper for the generation of nitrous air, I firſt aſcribed this circumſtance to that property of this metal, by which it burns with a green flame; but I was preſently ſatisfied that it muſt ariſe from the ſpirit of nitre, for the effect is the very ſame from whichever of the metals the nitrous air is extracted, all of which I tried for this purpoſe, even ſilver and gold. A mixture of oil of vitriol and ſpirit of nitre in equal proportions diſſolved iron, and the produce was nitrous air; but a leſs degree of ſpirit of nitre in the mixture produced air that was inflammable, and which burned with a green flame. It alſo tinged common air a little red, and diminiſhed it, though not much.

The diminution of common air by a mixture of nitrous air, is not ſo extraordinary as the diminution which nitrous air itſelf is ſubject to from a mixture of iron filings and brimſtone, made into a paſte with water. This mixture, as I have already obſerved, diminiſhes common air between one fifth and one fourth, but has no ſuch effect upon any kind of air that has been diminiſhed, and rendered noxious by any other proceſs; but when it is put to a quantity of nitrous air, it diminiſhes it ſo much, that no more than one fourth of the original quantity will be left. The effect of this proceſs is generally perceived in five or ſix hours, about which time the viſible effervefcence of the mixture begins; and in a very ſhort time it advances ſo rapidly, that in about an hour almoſt the whole effect will have taken place. If it be ſuffered to ſtand a day or two longer, the air will ſtill be diminiſhed farther, but only a very little farther, in proportion to the firſt diminution. The glaſs jar, in which the air and this mixture have been confined, has generally been ſo much heated in this proceſs, that I have not been able to touch it.

Nitrous air thus diminiſhed has not the peculiar ſmell of nitrous air, but ſmells juſt like common air in which the ſame mixture has ſtood; and it is not capable of being diminiſhed any farther, by a freſh mixture of iron and brimſtone.

Common air ſaturated with nitrous air is alſo no farther diminiſhed by this mixture of iron filings and brimſtone, though the mixture ferments with great heat, and ſwells very much in it.

Plants die very ſoon, both in nitrous air, and alſo in common air ſaturated with nitrous air, but eſpecially in the former.

Neither nitrous air, nor common air ſaturated with nitrous air, differs in ſpecific gravity from common air, or, at leaſt, ſo little, that I could not be ſure of it, ſometimes about three pints of it ſeeming to be about half a grain heavier, and at other times as much lighter than common air.

Having, among other kinds of air, expoſed a quantity of nitrous air, to water out of which the air had been well boiled, in the experiment to which I have more than once referred, as having been the occaſion of ſeveral new and important obſervations, I found that $\frac{1}{20}$ of the whole was abſorbed. Perceiving, to my great ſurprize, that ſo very great a proportion of this kind of air was miſcible with water, I immediately began to agitate a conſiderable quantity of it, in a jar ſtanding in a trough of the ſame kind of water; and with about four times as much agitation as fixed air requires, it was ſo far abſorbed by the water, that only about one fifth remained. This remainder extinguiſhed flame, and was noxious to animals. Afterwards I diminiſhed a pretty large quantity of it to one eighth of its original bulk, and the remainder ſtill retained much of its peculiar ſmell, and diminiſhed common air a little. A mouſe alſo died in it, but not ſo ſuddenly as it would have done in pure nitrous air. In this operation the peculiar ſmell of nitrous air is very manifeſt, the water being firſt impregnated with the air, and then tranſmitting it to the common atmoſphere.

This experiment gave me the hint of impregnating water with nitrous air, in the manner in which I had before done it with fixed air; and I preſently found that diſtilled water would imbibe about one tenth of its bulk of this kind of air, and

that

that it acquired a remarkably acid and aſtringent taſte from it. The ſmell of water thus impregnated is at firſt peculiarly pungent. I did not chuſe to ſwallow any of it, though, for any thing that I know, it may be perfectly innocent, and perhaps, in ſome caſes, ſalutary.

This kind of air is retained very obſtinately by water. In an exhauſted receiver a quantity of water thus ſaturated emitted a whitiſh fume, ſuch as ſometimes iſſues from bubbles of this air when it is firſt generated, and alſo ſome air bubbles; but though it was ſuffered to ſtand a long time in this ſituation, it ſtill retained its peculiar taſte; but when it had ſtood all night pretty near the fire, the water was become quite vapid, and had depoſited a filmy kind of matter, of which I had often collected a conſiderable quantity from the trough in which jars containing this air had ſtood. This I ſuppoſe to be a precipitate of the metal by the ſolution of which the nitrous air was generated. I have not given ſo much attention to it as to know, with certainty, in what circumſtances this depoſit is made, any more than I do the matter depoſited from inflammable air abovementioned; for I cannot get it, at leaſt in any conſiderable quantity, when I pleaſe; whereas I have often found abundance of it, when I did not expect it at all.

The nitrous air with which I made the firſt impregnation of water was extracted from copper; but when I made the impregnation with air from quickſilver, the water had the very ſame taſte, though the matter depoſited from it ſeemed to be of a different kind; for it was whitiſh, whereas the other had a yellowiſh tinge. Except the firſt quantity of this impregnated water, I could never deprive any more that I made of its peculiar taſte. I have even let ſome of it ſtand more than a week, in phials with their mouths open, and ſometimes very near the fire, without producing any alteration in it.

Whether any of the ſpirit of nitre be properly contained in the nitrous air, and be mixed with the water in this operation, I have not yet endeavoured to determine. This, however, may probably be the caſe, as the ſpirit of nitre is in a conſiderable degree volatile.

It will perhaps be thought, that the moſt uſeful, if not the moſt remarkable, of all the properties of this extraordinary kind of air, is its power of preſerving animal ſubſtances from putrefaction, and of reſtoring thoſe that are already putrid, which it poſſeſſes in a far greater degree than fixed air. My firſt obſervation of this was altogether caſual. Having found nitrous air to ſuffer ſo great a diminution as I have already mentioned by a mixture of iron filings and brimſtone, I was willing to try whether it would be equally diminiſhed by other cauſes of the diminution of common air, eſpecially by putrefaction; and for this purpoſe I put a dead mouſe into a quantity of it, and placed it near the fire, where the tendency to putrefaction was very great. In this caſe there was a conſiderable diminution, *viz.* from $5\frac{1}{4}$ to $3\frac{1}{4}$; but not ſo great as I had expected, the antiſeptic power of the nitrous air having checked

the

the tendency to putrefaction; for when, after a week, I took the mouſe out, I perceived, to my very great ſurprize, that it had no offenſive ſmell.

Upon this I took two other mice, one of them juſt killed, and the other ſoft and putrid, and put them both into the ſame jar of nitrous air, ſtanding in the uſual temperature of the weather, in the months of July and Auguſt of 1772; and after 25 days, having obſerved that there was little or no change in the quantity of the air, I took the mice out; and, examining them, found them both perfectly ſweet, even when cut through in all places. That which had been put into the air when juſt dead was quite firm; and the fleſh of the other, which had been putrid and ſoft, was ſtill ſoft, but perfectly ſweet.

In order to compare the antiſeptic power of this kind of air with that of fixed air, I examined a mouſe which I had incloſed in a phial full of fixed air, as pure as I could make it, and which I had corked very cloſe; but upon opening this phial in water, about a month after, I perceived that a large quantity of putrid effluvium had been generated; for it ruſhed with violence out of the phial; and the ſmell that came from it, the moment the cork was taken out, was inſufferably offenſive. Indeed Dr. Macbride ſays, that he could only reſtore very thin pieces of putrid fleſh by means of fixed air. Perhaps the antiſeptic power of theſe kinds of air may be in proportion to their acidity. If a little pains were taken with this ſubject, this remarkable antiſeptic power of nitrous air might poſſibly be applied to various uſes, perhaps to the

preſervation

preſervation of the more delicate birds, fiſhes, fruits, &c. mixing it in different proportions with common or fixed air. Of this property of nitrous air anatomiſts may perhaps avail themſelves, as animal ſubſtances may by this means be preſerved in their natural ſoft ſtate; but how long it will anſwer for this purpoſe, experience only can ſhew.

I calcined lead and tin in the manner hereafter deſcribed in a quantity of nitrous air, but with very little ſenſible effect; which rather ſurprized me; as, from the reſult of the experiment with the iron filings and brimſtone, I had expected a very great diminution of the nitrous air by this proceſs, the mixture of iron filings and brimſtone, and the calcination of metals, having the ſame effect upon common air, both of them diminiſhing it in nearly the ſame proportion.

Nitrous air is procured from all the proper metals by ſpirit of nitre, except lead, and from all the ſemi-metals that I have tried, except zinc. For this purpoſe I have uſed biſmuth and nickel, with ſpirit of nitre only, and regulus of antimony and platina, with aqua regia.

I got little or no air from lead by ſpirit of nitre, and have not yet made any experiments to aſcertain the nature of this ſolution. With zinc I have taken a little pains.

Four penny weights and ſeventeen grains of zinc diſſolved in ſpirit of nitre, to which as much water was added, yielded about twelve ounce meaſures of air, which had, in ſome degree, the properties of nitrous air, making a ſlight effervеſcence with common air, and diminiſhing it about as much as nitrous

trous air, which had been itſelf diminiſhed one half by waſhing in water. The ſmell of them both was alſo the ſame; ſo that I concluded it to be the ſame thing, that part of the nitrous air which is imbibed by water being retained in this ſolution.

In order to diſcover whether this was the caſe, I made the ſolution boil in a ſand heat. Some air came from it in this ſtate, which ſeemed to be the ſame thing, as nitrous air diminiſhed about one ſixth, or one eighth, by waſhing in water. When the fluid part was evaporated, there remained a brown fixed ſubſtance, which was obſerved by Mr. Hellot, who deſcribes it, Ac. Par. 1735, M. p. 35. A part of this I threw into a ſmall red hot crucible; and covering it immediately with a receiver, ſtanding in water, I obſerved that very denſe red fumes roſe from it, and filled the receiver. This redneſs continued about as long as that which is occaſioned by a mixture of nitrous and common air; the air was alſo conſiderably diminiſhed within the receiver. This ſubſtance, therefore, muſt certainly have contained within it the very ſame thing, or principle, on which the peculiar properties of nitrous air depend. It is remarkable, however, that though the air within the receiver was diminiſhed about one fifth by this proceſs, it was itſelf as much affected with a mixture of nitrous air, as common air is, and a candle burnt in it very well. This may perhaps be attributed to ſome effect of the ſpirit of nitre, in the compoſition of that brown ſubſtance.

Nitrous air, I find, will be conſiderably diminiſhed in its bulk by ſtanding a long time in water,

ter, about as much as inflammable air is diminiſhed in the ſame circumſtances. For this purpoſe I kept for ſome months a quart bottle full of each of theſe kinds of air; but as different quantities of inflammable air vary very much in this reſpect, it is not improbable but that nitrous air may vary alſo.

From one trial that I made, I conclude that nitrous air may be kept in a bladder much better than moſt other kinds of air. The air to which I refer was kept about a fortnight in a bladder, through which the peculiar ſmell of the nitrous air was very ſenſible for ſeveral days. In a day or two the bladder became red, and was much contracted in its dimenſions. The air within it had loſt very little of its peculiar property of diminiſhing common air.

I did not endeavour to aſcertain the exact quantity of nitrous air produced from given quantities of all the metals which yield it; but the few obſervations which I did make for this purpoſe I ſhall recite in this place:

dwt.	gr.			
6	0	of ſilver yielded	17½	ounce meaſures
5	19	of quickſilver	4½	
1	2½	of copper	14½	
2	0	of braſs	21	
0	20	of iron	16	
1	5	of biſmuth	6	
0	12	of nickel	4	

VII. Of

VII.

Of Air infected with the fumes of burning charcoal.

Air infected with the fumes of burning charcoal is well known to be noxious; and the Honourable Mr. Cavendiſh favoured me with an account of ſome experiments of his, in which a quantity of common air was reduced from 180 to 162 ounce meaſures, by paſſing through a red-hot iron tube filled with the duſt of charcoal. This diminution he aſcribed to ſuch a deſtruction of common air as Dr. Hales imagined to be the conſequence of burning. Mr. Cavendiſh alſo obſerved, that there had been a generation of fixed air in this proceſs, but that it was abſorbed by ſope leys. This experiment I alſo repeated, with a ſmall variation of circumſtances, and with nearly the ſame reſult.

Afterwards, I endeavoured to aſcertain, by what appears to me to be an eaſier and a more certain method, in what manner air is affected with the fumes of charcoal, viz. by ſuſpending bits of charcoal within glaſs veſſels, filled to a certain height with water, and ſtanding inverted in another veſſel of water, while I threw the focus of a burning mirror, or lens, upon them. In this manner I diminiſhed a given quantity of air one fifth, which is nearly in the ſame proportion with other diminutions of air.

Some fixed air ſeems to be contained in charcoal, and to be ſet looſe from it by this proceſs; for if I made uſe of lime-water, it never failed to become

 turbid,

turbid, presently after the heat was applied. This was the case with whatever degree of heat the charcoal had been made. If, however, the charcoal had not been made with a very considerable degree of heat, there never failed to be a permanent addition of inflammable air produced; which agrees with what I observed before, that, in converting dry wood into charcoal, the greatest part is changed into inflammable air. I have sometimes found, that charcoal which was made with the most intense heat of a smith's fire, which vitrified part of a common crucible in which the charcoal was confined, and which had been continued above half an hour, did not diminish the air in which the focus of a burning mirror was thrown upon it; a quantity of inflammable air equal to the diminution of the common air being generated in the process; whereas, at other times, I have not perceived that there was any generation of inflammable air, but a perfect diminution of common air, when the charcoal had been made with a much less degree of heat. This subject deserves to be farther investigated.

To make the preceding experiment with still more accuracy, I repeated it in quicksilver; when I perceived that there was a small increase of the quantity of air, from a generation either of fixed or inflammable air, but I suppose of the former. Thus it stood without any alteration a whole night, and part of the following day; when lime-water, being admitted to it, it presently became turbid, and, after some time, the whole quantity of air, which was about four ounce measures, was diminished one fifth, as before. In this case, I carefully weighed the piece of charcoal, which was exactly two grains, and could not find that

that it was ſenſibly diminiſhed in weight by the operation.

Air thus diminiſhed by the fumes of burning charcoal not only extinguiſhes flame, but is in the higheſt degree noxious to animals; it makes no efferveſcence with nitrous air, and is incapable of being diminiſhed any farther by the fumes of more charcoal, by a mixture of iron filings and brimſtone, or by any other cauſe of the diminution of air that I am acquainted with.

This obſervation, which reſpects all other kinds of diminiſhed air, proves that Dr. Hales was miſtaken in his notion of the abſorption of air in thoſe circumſtances in which he obſerved it. For he ſuppoſed that the remainder was, in all caſes, of the ſame nature with that which had been abſorbed, and that the operation of the ſame cauſe would not have failed to produce a farther diminution; whereas all my obſervations not only ſhew that air, which has once been fully diminiſhed by any cauſe whatever, is not only incapable of any farther diminution, either from the ſame or from any other cauſe, but that it has likewiſe acquired new properties, moſt remarkably different from thoſe which it had before, and that they are, in a great meaſure, the ſame in all the caſes. Theſe circumſtances give reaſon to ſuſpect, that the cauſe of diminution is, in reality, the ſame in all the caſes. What this cauſe is, may, perhaps, appear in the next courſe of obſervations.

VIII.

Of the effect of the calcination of metals, and of the effluvia of paint made with white-lead and oil, on Air.

Having been led to ſuſpect, from the experiments which I had made with charcoal, that the diminution of air in that caſe, and perhaps in other caſes alſo, was, in ſome way or other, the conſequence of its having more than its uſual quantity of phlogiſton, it occurred to me, that the calcination of metals, which are generally ſuppoſed to conſiſt of nothing but a metallic earth united to phlogiſton, would tend to aſcertain the fact, and be a kind of *experimentum crucis* in the caſe. Accordingly, I ſuſpended pieces of lead and tin in given quantities of air, in the ſame manner as I had before treated the charcoal; and throwing the focus of a burning mirror or lens upon them, in ſuch a manner as to make them fume copiouſly, I preſently perceived a diminution of the air. In the firſt trial that I made, I reduced four ounce meaſures of air to three, which is the greateſt diminution of common air that I had ever obſerved before, and which I account for, by ſuppoſing that, in other caſes, there was not only a cauſe of diminution, but cauſes of addition alſo, either of fixed or inflammable air, or ſome other permanently elaſtic matter, but that, the effect of the calcination of metals being ſimply the eſcape of phlogiſton, the cauſe of diminution was alone and uncontrouled.

The

The air, which I had thus diminifhed by calcination of lead, I transferred into another clean phial, but found that the calcination of more lead in it had no farther effect upon it. This air alfo, like that which had been infected with the fumes of charcoal, was in the higheft degree noxious, made no effervefcence with nitrous air, was no farther diminifhed by the mixture of iron filings and brimftone, and was not only rendered innoxious, but alfo recovered, in a great meafure, the other properties of common air, by wafhing in water.

It might be fufpected that the noxious quality of the air in which lead was calcined, might be owing to fome fumes peculiar to that metal; but I found no fenfible difference between the properties of this air, and that in which tin was calcined.

The water over which metals are calcined acquires a yellowifh tinge, and an exceedingly pungent fmell and tafte, pretty much, as near as I can recollect, for I did not compare them together, like that over which brimftone has been frequently burned. Alfo a thin and whitifh pellicle covered both the furface of the water, and likewife the fides of the phial in which the calcination was made, infomuch that, without frequently agitating the water, it grew fo opaque by this conftantly accumulating incruftation, that the fun beams could not be tranfmitted through it in a quantity fufficient to produce the calcination.

I imagined, however, that, even when this air was transferred into a clean phial, the metals were not fo eafily melted or calcined as they were in frefh air; for the air being once fully faturated with phlogifton, may not fo readily admit any more, though it be only to tranſmit it to the water. I alſo ſuſpected that metals were not eaſily melted or calcined in inflammable, fixed, or nitrous, air, or any kind of diminiſhed air. None of theſe kinds of air ſuffered any change by this operation; nor was there any precipitation of lime, when charcoal was heated in any of theſe kinds of air ſtanding in lime-water.

Query. May not water impregnated with phlogiſton from calcined metals, or by any other method, be of ſome uſe in medicine? The effect of this impregnation is exceedingly remarkable; but the principle with which it is impregnated is volatile, and entirely eſcapes in a day or two, if the ſurface of the water be expoſed to the common atmoſphere.

It ſhould ſeem that phlogiſton is retained more obſtinately by charcoal than it is by lead or tin; for when any given quantity of air is fully ſaturated with phlogiſton from charcoal, no heat that I have yet applied has been able to produce any more effect upon it; whereas, in the ſame circumſtances, lead and tin may ſtill be calcined. The air, indeed, can take no more; but the water receives it, and the ſides of the phial alſo receive an addition of incruſtation. This is a white powdery ſubſtance, and well deſerves to be examined. I ſhall endeavour to do it at my leiſure.

Lime-water never became turbid by the calcination of metals over it; but the colour, ſmell, and taſte of the water was always changed, and the ſurface of it became covered with a yellow pellicle, as before.

When this proceſs was made in quickſilver, the air was diminiſhed only one fifth; and upon water being admitted to it, no more was abſorbed; which is an effect ſimilar to that of a mixture of nitrous and common air, which was mentioned before.

The preceding experiments on the calcination of metals ſuggeſted to me a method of explaining the cauſe of the miſchief which is known to ariſe from freſh paint, made with white lead (which I ſuppoſe is an imperfect calx of lead) and oil. To verify my hypotheſis, I firſt put a ſmall pot full of this kind of paint, and afterwards (which anſwered much better, by expoſing a greater ſurface of the paint) I daubed ſeveral pieces of paper with it, and put them under a receiver, and obſerved, that in about twenty-four hours, the air was diminiſhed between one fifth and one fourth, for I did not meaſure it very exactly. This air alſo was, as I expected to find it, in the higheſt degree, noxious; it did not effervеſce with nitrous air, it was no farther diminiſhed by a mixture of iron filings and brimſtone, and was made wholeſome by agitation in water deprived of all air.

I think it appears pretty evident, from the preceding experiments on the calcination of metals, that air is ſome way or other diminiſhed in conſequence of being highly charged with phlogiſton, and that agitation in water reſtores it, by imbibing a great part of the phlogiſtic matter. That water has a conſiderable affinity with phlogiſton, is evident from the ſtrong impregnation which it receives from it. May not plants alſo reſtore air diminiſhed by putrefaction, by abſorbing part of the phlogiſton with which it is loaded? The greater part of a dry plant, as well as of a dry animal ſubſtance, conſiſts of inflammable air, or ſomething that is capable of being converted

into

into inflammable air; and it ſeems to be as probable that this phlogiſtic matter may have been imbibed by the roots and leaves of plants, and afterwards incorporated into their ſubſtance, as that it is altogether produced by the power of vegetation. May not this phlogiſtic matter be even the moſt eſſential part of the food and ſupport of both vegetable and animal bodies?

In the experiments with metals, the diminution of air ſeems to be the conſequence of nothing but a ſaturation with phlogiſton; and in all the other caſes of the diminution of air, I do not ſee but that it may be effected by the ſame means. When a vegetable or animal ſubſtance is diſſolved by putrefaction, the eſcape of the phlogiſtic matter (which, together with all its other conſtituent parts, is then let looſe from it) may be the circumſtance that produces the diminution of the air in which it putrefies. It is highly improbable that what remains after an animal body has been thoroughly diſſolved by putrefaction, ſhould yield ſo great a quantity of inflammable air, as the dried animal ſubſtance would have done. Of this I have not made an actual trial, though I have often thought of doing it, and ſtill intend to do it; but I think there can be no doubt of the reſult. Again, the iron, by its fermentation with brimſtone and water, is evidently reduced to a calx, ſo that phlogiſton muſt have eſcaped from it. Phlogiſton alſo muſt evidently be ſet looſe by the ignition of charcoal, and is not improbably the matter which flies off from paint, compoſed of white lead and oil. Laſtly, ſince ſpirit of nitre is known to have a very remarkable affinity with phlogiſton, it is far from

being

being improbable that nitrous air may alſo produce the ſame effect by the ſame means.

To this hypotheſis it may be objected, that, if diminiſhed air be air ſaturated with phlogiſton, it ought to be inflammable; but this by no means follows, ſince its inflammability may depend upon ſome particular mode of combination, or degree of affinity, with which we are not acquainted. Beſides, inflammable air ſeems to conſiſt of ſome other principle, or to have ſome other conſtituent part, beſides phlogiſton and common air, as is probable from that remarkable depoſit, which, as I have obſerved, is made by inflammable air, both from iron and zinc.

It is not improbable, however, but that a greater degree of heat may inflame that air which extinguiſhes a common candle, if it could be conveniently applied. Air that is inflammable, I obſerve, extinguiſhes red hot wood; and indeed inflammable ſubſtances can only be thoſe which, in a certain degree of heat, have a leſs affinity with the phlogiſton they contain, than the air, or ſome other contiguous ſubſtance, has with it; ſo that the phlogiſton only quits one ſubſtance, with which it was before combined, and enters another, with which it may be combined in a very different manner. This ſubſtance, however, whether it be air or any thing elſe, being now fully ſaturated with phlogiſton, and not being able to take any more, in the ſame circumſtances, muſt neceſſarily extinguiſh fire, and put a ſtop to the ignition of all other bodies, that is, to the farther eſcape of phlogiſton from them.

That plants reſtore noxious air, by imbibing the phlogiſton with which it is loaded, is very agreeable to the conjectures of Dr. Franklin, made many years ago, and expressed in the following extract from the last edition of his Letters, p. 346.

" I have been inclined to think that the fluid *fire*, " as well as the fluid *air*, is attracted by plants in " their growth, and becomes consolidated with the " other materials of which they are formed, and " makes a great part of their substance; that, when " they come to be digested, and to suffer in the " vessels a kind of fermentation, part of the fire, as " well as part of the air, recovers its fluid active state " again, and diffuses itself in the body, digesting and " separating it; that the fire so reproduced, by di- " gestion and separation, continually leaving the " body, its place is supplied by fresh quantities, " arising from the continual separation; that what- " ever quickens the motion of the fluids in an ani- " mal quickens the separation, and re-produces " more of the fire, as exercise; that all the fire " emitted by wood, and other combustibles, when " burning, existed in them before, in a solid state, " being only discovered when separating; that some " fossils, as sulphur, sea-coal, &c. contain a great " deal of solid fire; and that, in short, what escapes " and is dissipated in the burning of bodies, besides " water and earth, is generally the air, and fire, " that before made parts of the solid."

IX.

Of Air procured by means of spirit of salt.

Being very much struck with the result of an experiment of the Hon. Mr. Cavendish, related Phil.

Trans.

Tranſ. Vol. LVI. p. 157. by which, though, he ſays, he was not able to get any inflammable air from copper, by means of ſpirit of ſalt, he got a much more remarkable kind of air, *viz.* one that loſt its elaſticity by coming into contact with water, I was exceedingly deſirous of making myſelf acquainted with it. On this account, I began with making the experiment in quickſilver, which I never failed to do in any caſe in which I ſuſpected that air might either be abſorbed by water, or be in any other manner affected by it; and by this means I preſently got a much more diſtinct idea of the nature and effects of this curious ſolution.

Having put ſome copper filings into a ſmall phial, with a quantity of ſpirit of ſalt; and making the air, which was generated in great plenty, on the application of heat, aſcend into a tall glaſs veſſel full of quickſilver, and ſtanding in quickſilver, the whole produce continued a conſiderable time without any change of dimenſions. I then introduced a ſmall quantity of water to it, when about three fourths of it (the whole being about four ounce meaſures) preſently, but gradually, diſappeared, the quickſilver riſing in the veſſel. I then introduced a conſiderable quantity of water; but there was no farther diminution of the air, and the remainder I found to be inflammable.

Having frequently continued this proceſs a long time after the admiſſion of the water, I was much amuſed with obſerving the large bubbles of the newly generated air, which came through the quickſilver, the ſudden diminution of them when they came to the water, and the very ſmall bubbles which went through the water. They made, however, a continual, though slow, increase of inflammable air.

Fixed air, being admitted to the whole produce of this air from copper, had no sensible effect upon it. Upon the admission of water, a great part of the mixture, which, no doubt, was the most subtle kind of air from the copper, presently disappeared; another part, which I suppose to have been the fixed air, was absorbed slowly; and in this particular case the very small permanent residuum did not take fire; but it is very possible that it might have done so, if the quantity had been greater.

Lime-water being admitted to the whole produce of air from copper became white; but this I suspect to have arisen from some other circumstance than the precipitation of the lime which it contained.

The solution of lead in the marine acid is attended with the very same phænomena as the solution of copper in the same acid; about three fourths of the generated air disappearing on the contact of water, and the remainder being inflammable.

The solutions of iron, tin, and zinc, in the marine acid, were all attended with the same phænomena as the solutions of copper and lead, but in a less degree; for in iron one eighth, in tin one sixth, and in zinc one tenth of the generated air disappeared on the contact with water. The remainder of the air from iron, in this case, burned with a green, or very light blue flame.

I had always thought it something extraordinary that a species of air should lose its elasticity by the mere contact of any thing, and from the first suspected that it must have been imbibed by the water

that

that was admitted to it; but ſo very great a quantity of this air diſappeared upon the admiſſion of a very ſmall quantity of water, that I could not help concluding that appearances favoured the former hypotheſis. I found, however, that when I admitted a much ſmaller quantity of water, confined in a narrow glaſs tube, a part only of the air diſappeared, and that very ſlowly, and that more of it vaniſhed upon the admiſſion of more water. This obſervation put it beyond a doubt, that this air was properly imbibed by the water, which, being once fully ſaturated with it, was not capable of receiving any more. The water thus impregnated taſted very acid, even when it was much diluted with other water, through which the tube containing it was drawn. It even diſſolved iron very faſt, and generated inflammable air. This laſt obſervation, together with another which immediately follows, led me to the diſcovery of the true nature of this remarkable kind of air, as it had hitherto been called.

Happening, at one time, to uſe a good deal of copper and a ſmall quantity of ſpirit of ſalt, in the generation of this kind of air, I was ſurprized to find that air was produced long after, I could not but think that the acid muſt have been ſaturated with the metal; and I alſo found that the proportion of inflammable air to that which was abſorbed by the water continually diminiſhed, till, inſtead of being one fourth of the whole as I had firſt obſerved, it was not ſo much as one twentieth. Upon this, I concluded that this ſubtle air did not ariſe from the copper, but from the ſpirit of ſalt; and preſently making the experiment with the acid only, without any cop-

4 per,

per, or metal of any kind, this air was immediately produced in as great plenty as before; ſo that this remarkable kind of air is, in fact, nothing more than the vapour, or fumes of ſpirit of ſalt, which appear to be of ſuch a nature, that they are not liable to be condenſed by cold, like the vapour of water, and other fluids. This vapour, however, ſeems to loſe its elaſticity, in ſome meaſure, gradually, unleſs it ſhould be thought to be affected by the quickſilver, with which it is in contact; for it was always diminiſhed, more or leſs, by ſtanding.

This elaſtic acid vapour extinguiſhes flame, and is much heavier than common air; but how much heavier, will not be eaſy to aſcertain. A cylindrical glaſs veſſel, about three fourths of an inch in diameter, and four inches deep, being filled with it, and turned upſide down, a lighted candle may be let down into it more than twenty times before it will burn at the bottom. It is pleaſing to obſerve the colour of the flame in this experiment; for both before the candle goes out, and alſo when it is firſt lighted again, it burns with a beautifully green, or rather light blue flame, ſuch as is ſeen when common ſalt is thrown into the fire.

When this elaſtic vapour is all expelled from any quantity of ſpirit of ſalt, which is eaſily perceived by the vapour being condenſed by cold, the remainder is a very weak acid, barely capable of diſſolving iron.

Being now in the poſſeſſion of a new ſubject of experiments, *viz.* an elaſtic acid vapour, in the form of a permanent air, eaſily procured, and effectually confined by glaſs and quickſilver, with which

which it did not ſeem to have any affinity; I immediately began to introduce a variety of ſubſtances to it, in order to aſcertain its peculiar properties and affinities, and alſo the properties of thoſe other bodies with reſpect to it.

Beginning with water, which, from preceding obſervations, I knew would imbibe it, and become impregnated with it; I found that 2½ grains of rain water abſorbed three ounce meaſures of this vapour, after which it was increaſed one third in its bulk, and weighed twice as much as before; ſo that this concentrated vapour ſeems to be twice as heavy as rain water. Water impregnated with it makes the ſtrongeſt ſpirit of ſalt that I have ſeen, diſſolving iron with the moſt rapidity. Conſequently, two thirds of the beſt ſpirit of ſalt is nothing more than mere phlegm or water.

Iron filings, being admitted to this vapour, were diſſolved by it pretty faſt, half of the vapour diſappearing, and the other half becoming inflammable air, not abſorbed by water. Putting chalk to it, fixed air was produced.

I had not introduced many ſubſtances to this vapour, before I diſcovered that it had an affinity with phlogiſton, ſo that it would deprive other ſubſtances of it, and form with it ſuch an union as conſtitutes inflammable air; which ſeems to ſhew, that inflammable air univerſally conſiſts of the union of ſome acid vapour with phlogiſton.

Inflammable air was produced, when to this vapour I put ſpirit of wine, oil of olives, oil of turpentine, charcoal, phoſphorus, bees-wax, and even ſulphur. This laſt obſervation, I own, ſur-

prized

prized me; for, the marine acid being reckoned the weakeſt of the three mineral acids, I did not think that it had been capable of diſlodging the oil of vitriol from this ſubſtance; but I found that it had the very ſame effect both upon alum and nitre; the vitriolic acid in the former caſe, and the nitrous in the latter, giving place to the ſtronger vapour of ſpirit of ſalt.

The ruſt of iron, and the precipitate of nitrous air made from copper, alſo imbibed this vapour very faſt, and the little that remained of it was inflammable air; which proves, that theſe calces contain phlogiſton. It ſeems alſo to be pretty evident, from this experiment, that the precipitate above-mentioned is a real calx of the metal, by the ſolution of which the nitrous air is generated.

As ſome remarkable circumſtances attend the abſorption of this vapour of ſpirit of ſalt, by the ſubſtances above-mentioned, I ſhall briefly mention them.

Spirit of wine abſorbs this vapour as readily as water itſelf, and is increaſed in bulk by that means. Alſo, when it is ſaturated, it diſſolves iron with as much rapidity, and ſtill continues inflammable.

Oil of olives abſorbs this vapour very ſlowly, and, at the ſame time, it turns almoſt black, and becomes glutinous. It is alſo leſs miſcible with water, and acquires a very diſagreeable ſmell. By continuing upon the ſurface of the water, it became white, and its offenſive ſmell went off in a few days.

Oil of turpentine abſorbed this vapour very faſt, turning brown, and almoſt black. No inflammable air was formed, till I raiſed more of the vapour than the

the oil was able to abſorb, and let it ſtand a conſiderable time; and ſtill the air was but weakly inflammable. The ſame was the caſe with the oil of olives, in the laſt mentioned experiment; and it ſeems to be probable, that, the longer this acid vapour had continued in contact with the oil, the more phlogiſton it would have extracted from it. It is not improbable, but that, in the intermediate ſtate, before it becomes inflammable air, it may be nearly of the nature of common air.

Bees-wax abſorbed this vapour very ſlowly. About the bigneſs of a hazel-nut of the wax being put to three ounce meaſures of the vapour, the vapour was diminiſhed one half in two days, and, upon the admiſſion of water, half of the remainder alſo diſappeared. This air was ſtrongly inflammable.

Charcoal abſorbed this vapour very faſt. About one fourth of it was rendered immiſcible in water, and was but weakly inflammable.

A ſmall bit of phoſphorus, perhaps about half a grain, ſmoked, and gave light in the vapour of ſpirit of ſalt, juſt as it would have done in common air confined. It was not ſenſibly waſted after continuing about twelve hours in that ſtate, and the bulk of the vapour was very little diminiſhed. Water being admitted to it abſorbed it as before, except about one fifth of the whole, which was but weakly inflammable.

Putting ſeveral pieces of ſulphur to this vapour, it was abſorbed but ſlowly. In about twenty-four hours about one fifth of the quantity had diſappeared; and water being admitted to the remainder, very little more was abſorbed. The remainder was inflammable, and burned with a blue flame.

Nowithſtanding the affinity which this vapour of ſpirit of ſalt appears to have with phlogiſton, it is not capable of depriving all bodies of it. I found that dry wood, cruſts of bread, and raw fleſh, very readily imbibed this acid vapour, but did not part with any of their phlogiſton to it. All theſe ſubſtances turned very brown, after they had been ſome time expoſed to this vapour, and taſted very ſtrongly of the acid when they were taken out; but the fleſh, when waſhed in water, became very white, and the fibres eaſily ſeparated from one another, even more than they would have done if it had been boiled or roaſted.

When I put a piece of ſaltpetre to this vapour, it was preſently ſurrounded with a white fume, which ſoon filled the whole veſſel, exactly like the fume which burſts from the bubbles of nitrous air, when it is generated by a vigorous fermentation, and ſuch as is ſeen when nitrous air is mixed with this vapour of ſpirit of ſalt. In about a minute, the whole quantity of vapour was abſorbed, except a very ſmall quantity, which might be the common air that had lodged upon the ſurface of the ſpirit of ſalt within the phial.

A piece of alum expoſed to this vapour turned yellow, abſorbed it as faſt as the ſaltpetre had done, and was reduced by it to the form of a powder. The ſurface both of the nitre and alum was, I doubt not, changed into common ſalt, by this proceſs. Common ſalt, as might be expected, had no effect whatever on this vapour.

From

From confidering the affinity which this vapour has with phlogifton, I was induced to try the effect of a mixture of it with nitrous air. Accordingly, to two parts of this vapour, I put one part of nitrous air, and, in about twenty-four hours, the whole was diminifhed to fomething lefs than the original quantity of the vapour, and was no farther diminifhed by the admiffion of water. Holding the flame of a candle over this air, the lower part of it burned green, but there was no fenfible explofion. At different times I collected 2¾ ounce meafures of this mixture of air; but, upon agitating it in rain-water, it was prefently diminifhed to 1½ ounce meafures. In this ftate it effervefced with nitrous air, and was confiderably diminifhed by it, but not fo much as common air. Some allowance, no doubt, muft be made for the fmall quantities of common air, which lodged on the top of my phials, when I raifed the fume from the fpirit of falt; but, from the precautions that I made ufe of, I think that very little is to be allowed to this circumftance; and, upon the whole, I am of opinion, that this experiment is an approach to the generation of common air, or air fit for refpiration.

I had alfo imagined, that if air diminifhed by the proceffes above-mentioned was affected in this manner, in confequence of its being faturated with phlogifton, a mixture of this vapour might imbibe that phlogifton, and render it wholefome again; but I put about one fourth of this vapour to a quantity of air in which metals had been calcined, without making any fenfible alteration in it. I do not, however, infer from this, that air is not diminifhed by means of phlogifton, fince the air, like fome other fubftances,

 may

may hold the phlogiſton too faſt, to be deprived of it by this acid vapour.

I ſhall conclude my account of theſe experiments with obſerving, that the electric ſpark is viſible in the vapour of ſpirit of ſalt, exactly as it is in common air; and though I kept making this ſpark a conſiderable time in a quantity of it, I did not perceive that any ſenſible alteration was made in it. A little inflammable air was produced, but not more than might have come from the two iron nails which I made uſe of in taking the ſparks.

X.

Miscellaneous Observations.

Many of the preceding obſervations relating to the vinous and putrefactive fermentations, I had the curioſity to endeavour to aſcertain in what manner the air would be affected by the acetous fermentation. For this purpoſe I incloſed a phial full of ſmall beer in a jar ſtanding in water, and obſerved that during the firſt two or three days there was an increaſe of the air in the jar, but from that time it gradually decreaſed, till at length there appeared to be a diminution of about $\frac{1}{10}$ of the whole quantity. During this time the whole ſurface of it was gradually covered with a ſcum, beautifully corrugated. After this there was an increaſe of the air till there was more than the original quantity; but this muſt have been fixed air, not incorporated with the reſt of the maſs; for, withdrawing the beer, which I found to be ſour, after it had ſtood 18 or 20 days under the jar, and

paſſing

paſſing the air ſeveral times through cold water, the original quantity was diminiſhed about $\frac{1}{9}$. In the remainder a candle would not burn, and a mouſe would have died preſently. The ſmell of this air was exceedingly pungent, but different from that of the putrid effluvium. A mouſe lived perfectly well in this air, thus affected with the acetous fermentation; after it had ſtood ſeveral days mixed with four times the quantity of fixed air.

All the kinds of factitious air on which I have yet made the experiment are highly noxious to animals, except that which is extracted from ſalt-petre, or alum; but in this even a candle burned juſt as in common air. In one quantity which I got from ſalt-petre a candle not only burned, but the flame was increaſed, and ſomething was heard like a hiſſing, ſimilar to the decrepitation of nitre in an open fire. This experiment was made when the air was freſh made, and while it probably contained ſome particles of nitre, which would have been depoſited afterwards. The air was extracted from theſe ſubſtances by putting them into a gun barrel, which was much corroded and ſoon ſpoiled by the experiment. What effect this circumſtance may have had upon the air I have not conſidered.

November 6, 1772, I had the curioſity to examine the ſtate of a quantity of this air, which had been extracted from ſalt-petre above a year, and which at firſt was perfectly wholeſome; when, to my very great ſurprize, I found that it was become, in the higheſt degree, noxious. It made no efferveſcence with nitrous air, and a mouſe died the moment it was put into it. I had not, however, waſhed it in rain water quite ten minutes

(and

(and perhaps leſs time would have been ſufficient) when I found, upon trial, that it was reſtored to its former perfectly wholeſome ſtate. It effervеſced with nitrous air as much as the beſt common air ever does, and even a candle burned in it very well, which I had never before obſerved of any kind of noxious air meliorated by agitation in water. This ſeries of facts, relating to air extracted from nitre, appear to me to be very extraordinary and important, and, in able hands, may lead to conſiderable diſcoveries.

There are many ſubſtances which impregnate the air in a very remarkable manner, but without making it noxious to animals. Among other things I tried volatile alkaline ſalts, and camphire, the latter of which I melted with a burning glaſs, in air incloſed in a phial. The mouſe which was put into this air ſneezed and coughed very much, eſpecially after it was taken out; but it preſently recovered, and did not appear to have been ſenſibly injured.

Having made ſeveral experiments with a mixture of iron filings and brimſtone, kneaded to a paſte with water, I had the curioſity to try what would be the effect of ſubſtituting braſs duſt in the place of the iron filings. The reſult was, that when this mixture had ſtood about three weeks, in a given quantity of air, it had turned black, but was not increaſed in bulk. The air alſo was neither ſenſibly increaſed nor decreaſed, but the nature of it was changed, for it extinguiſhed flame, it would have killed a mouſe preſently, and was not reſtored by fixed air, which had been mixed with it ſeveral days.

I have

I have frequently mentioned my having, at one time, expoſed equal quantities of different kinds of air in jars ſtanding in boiled water. The common air in this experiment was diminiſhed four ſevenths, and the remainder extinguiſhed flame. This experiment demonſtrates that water does not abſorb air equally, but that it decompoſes it, taking one part, and leaving the reſt. To be quite ſure of this fact, I agitated a quantity of common air in boiled water, and when I had reduced it from eleven ounce meaſures to ſeven, I found that it extinguiſhed a candle, but a mouſe lived in it very well. At another time a candle barely went out when the air was diminiſhed one third, and at other times I have found this effect take place at other very different degrees of diminution. This difference I attribute to the differences in the ſtate of the water with reſpect to the air contained in it; for ſometimes it had ſtood longer than at other times before I made uſe of it. I alſo uſed diſtilled water, rain water, and water out of which the air had been pumped, promiſcuouſly with rain water. I even doubt not but that, in a certain ſtate of the water, there might be no ſenſible difference in the bulk of the agitated air, and yet at the end of the proceſs it would extinguiſh a candle, air being ſupplied from the water in the place of that part of the common air which had been abſorbed.

It is certainly a little extraordinary that the very ſame proceſs ſhould ſo far mend putrid air, as to reduce it to the ſtandard of air in which candles have burned out; and yet that it ſhould ſo far injure common and wholeſome air, as to reduce it to about

the

the ſame ſtandard: but ſo the fact certainly is. If air extinguiſh flame in conſequence of its being previouſly ſaturated with phlogiſton, it muſt, in this caſe, have been transferred from the water to the air.

To a quantity of common air, thus diminiſhed by agitation in water, till it extinguiſhed a candle, I put a plant, but it did not ſo far reſtore it as that a candle would burn in it again; which to me appeared not a little extraordinary, as it did not ſeem to be in a worſe ſtate than air in which candles had burned out, and which had never failed to be reſtored by the ſame means. I had no better ſucceſs with a quantity of permanent air; which I had collected from my pump water. Indeed theſe experiments were begun before I was acquainted with that property of nitrous air, which makes it ſo accurate a meaſure of the goodneſs of other kinds of air; and it might perhaps be rather too late in the year when I made the experiments. Having neglected theſe two jars of air, the plants died and putrefied in both of them; and then I found the air in them both to be highly noxious, and to make no effervеſcence with nitrous air.

I found that a pint of my pump water contains about one fourth of an ounce meaſure of air, one half of which was afterwards abſorbed by ſtanding in freſh pump water. A candle would not burn in the air, but a mouſe lived in it very well. Upon the whole, it ſeemed to be in about the ſame ſtate as air in which a candle had burned out.

I once imagined that, by mere ſtagnation, air might become unfit for reſpiration, or at leaſt for the burning of candles; but if this be the caſe, and the change be produced gradually, it muſt require a long time for the purpoſe. For on the 22d of September 1772, I examined a quantity of common air, which had been kept in a phial, without agitation, from May 1771, and found it to be in no reſpect worſe than freſh air, even by the teſt of the nitrous air.

The cryſtallization of nitre makes no ſenſible alteration in the air in which the proceſs is made. For this purpoſe I diſſolved as much nitre as a quantity of hot water would contain, and let it cool under a receiver, ſtanding in water.

November 6, 1772, a quantity of inflammable air, which, by long keeping, had come to extinguiſh flame, I obſerved to ſmell very much like common air in which a mixture of iron filings and brimſtone had ſtood. It was not, however, quite ſo ſtrong, but it was equally noxious.

Biſmuth and nickel are diſſolved in the marine acid with the application of a conſiderable degree of heat; but little or no air is got from either of them; but, what I thought a little remarkable, both of them ſmelled very much like Harrowgate water. This ſmell I have met with ſeveral times in the courſe of my experiments, and in proceſſes very different from one another.

As I generally made uſe of mice in the experiments which relate to reſpiration, and ſome perſons may chuſe to repeat them after me, and purſue them farther than I have done; it may be of uſe to them to be informed, that I kept them without any difficulty in glaſs receivers, open at the top and bottom, and having a quantity of paper, or tow, in the inſide, which ſhould be changed every three or four days; when it will be moſt convenient alſo to change the veſſel, and waſh it. But they muſt be kept in a pretty exact temperature, for either much heat or much cold kills them preſently. The place in which I have generally kept them is a ſhelf over the kitchin fire place, where, as it is uſual in Yorkſhire, the fire never goes out; ſo that the heat varies very little; and I find it to be at a medium about 70 degrees of Fahrenheit's thermometer. When they had been made to paſs through the water, as they neceſſarily muſt be, in order to a change of air, they require, and will bear a very conſiderable degree of heat, to warm and dry them.

I found, to my great ſurprize, in the courſe of theſe experiments, that mice will live intirely without water; for though I have kept ſome of them for three or four months, and have offered them water ſeveral times, they would never taſte it; and yet they continued in perfect health and vigour. Two or three of them will live very peaceably together in the ſame veſſel; though I had one inſtance of one mouſe tearing another almoſt in pieces, though there was plenty of proviſions for both of them.

The apparatus with which the principal of the preceding experiments were made is exceedingly ſimple, and cheap. The drawing annexed (TAB. IX.) exhibits a view of every thing that is moſt important in it.

A is

A is an oblong trough, about eight inches deep, kept nearly full of water, and B, B are jars ſtanding in it, about ten inches long, and two and a half wide; ſuch as I have generally uſed for electrical batteries.

C, C are flat ſtones, ſunk about an inch, or half an inch, under the water, on which veſſels of any kind may be conveniently placed, during a courſe of experiments.

D, D are pots nearly full of water, in which jars or phials, containing any kind of air, to which plants or any other ſubſtances may be expoſed, and having their mouths immerſed in water; ſo that the air in the inſide can have no communication with the external air.

E is a ſmall glaſs veſſel, of a convenient ſize for putting a mouſe into it, in order to try the wholeſomeneſs of any kind of air that it may contain.

F is a cylindrical glaſs veſſel, five inches in length, and one in diameter, very proper for trying whether any kind of air will admit a candle to burn in it. For this purpoſe a bit of wax candle, G, may be faſtened to the end of a wire, H, and turned up in ſuch a manner as to be let down into the veſſel with the flame upwards. The veſſel ſhould be kept carefully covered till the moment that the candle is admitted to it. In this manner I have frequently extinguiſhed a candle above twenty times in one of theſe veſſels full of air, though it is impoſſible to dip the candle into it, without giving the external air an opportunity of mixing with it, more or leſs.

I is a funnel of glaſs or tin, which is neceſſary for transferring air into veſſels which have narrow mouths.

K is a glaſs ſyphon, which is very uſeful for drawing air out of a veſſel which has its mouth immerſed in water, and thereby raiſing the water to whatever height may be moſt convenient. I do not think it by any means ſafe to depend upon a valve at the top of a veſſel, which Dr. Hales very often made uſe of; for, ſince my firſt diſappointments, I have never thought the communication between the external and internal air ſufficiently cut off, unleſs glaſs, or a body of water, or, in ſome caſes, quickſilver, have intervened between them.

L is a piece of a gun barrel, cloſed at one end, having the ſtem of a tobacco-pipe luted to the other. To the end of this pipe I ſometimes faſtened a flaccid bladder, in order to receive the air diſcharged from the ſubſtance contained in the barrel; but, when the air was generated ſlowly, I commonly contrived to put this end of the pipe under a veſſel full of water, and ſtanding with its mouth inverted in another veſſel of water, that the new air might have a more perfect ſeparation from the external air than a bladder could make.

M is a ſmall phial containing ſome mixture that will generate air. This air paſſes through a bent glaſs tube inſerted into the cork at one end, and going under the edge of the jar N at the other; the jar being placed with part of its mouth projecting beyond the flat ſtones C C for that purpoſe.

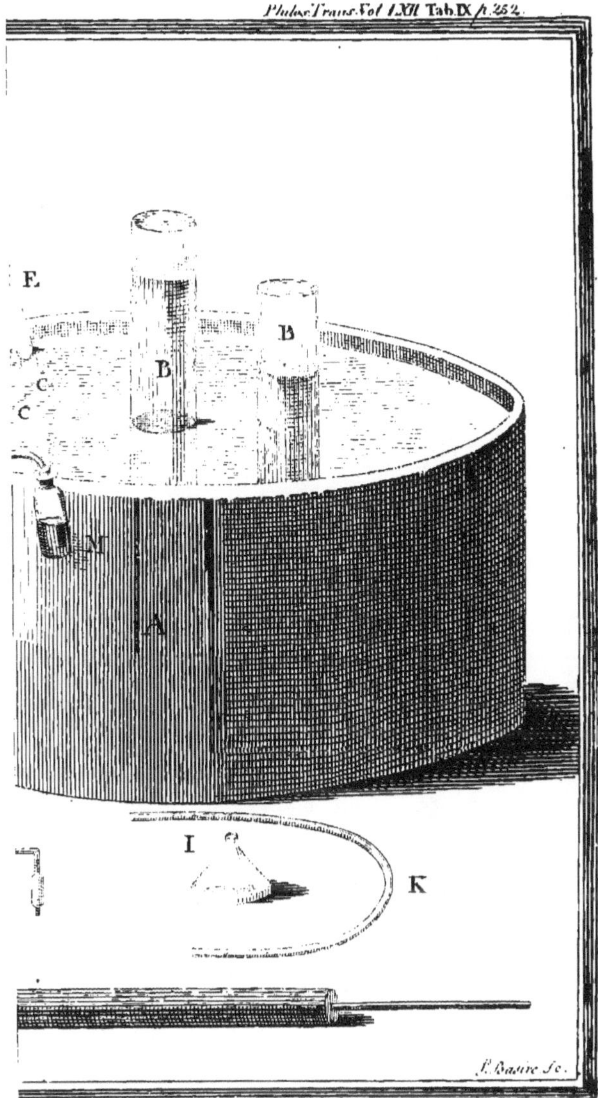
E
B
B
C
C
M
A
I
K
J. Basire Sc.

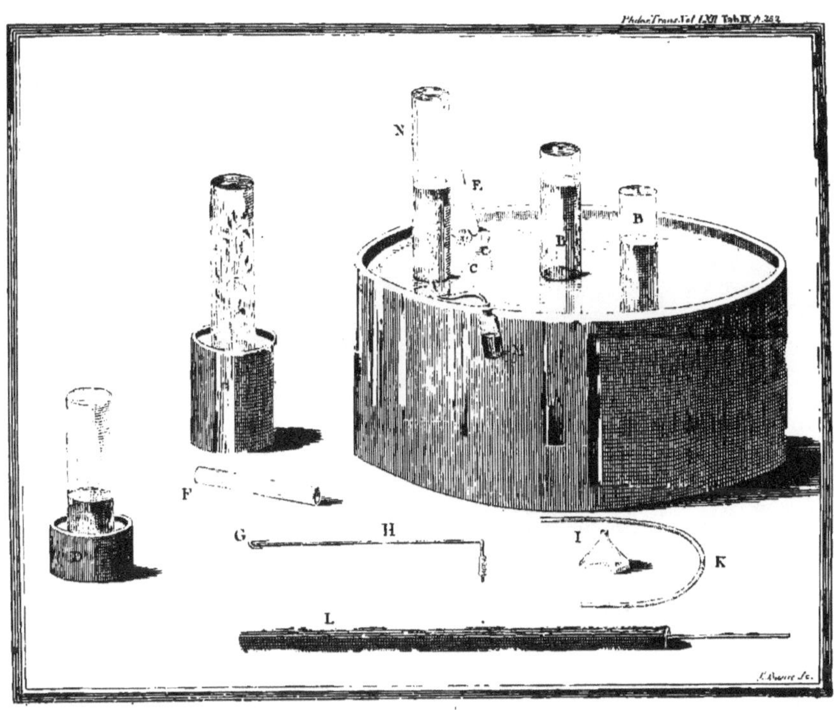
N
E
B
B
C
C
M
F
G
H
I
K
D
L

On the Ceiling of a CHOULTRY at VERDAPETTAH in the MADURAH COUNTRY taken the 8th of July 1764.

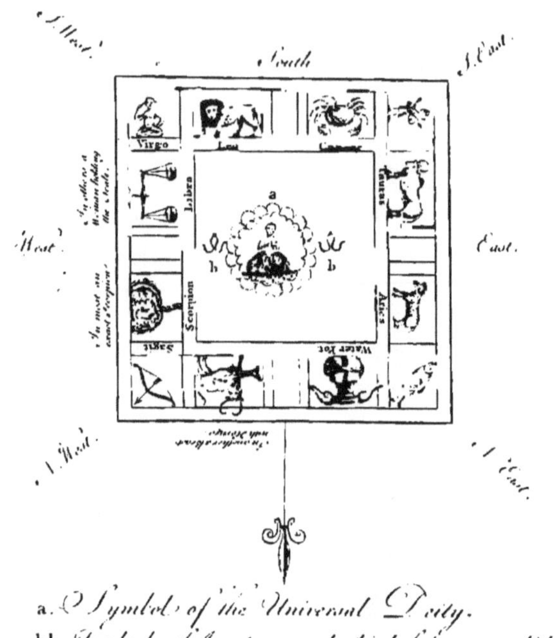

a. Symbol of the Universal Deity.

bb. Two hooks of Iron to suspend a kind of throne on which the Deity or Swamy often sat, when exhibited to the adorers.

On the Ceiling of a **CHOULT**

VERDAPETTAH *in the* **MADURAH C**

taken the 8th of July 1764.

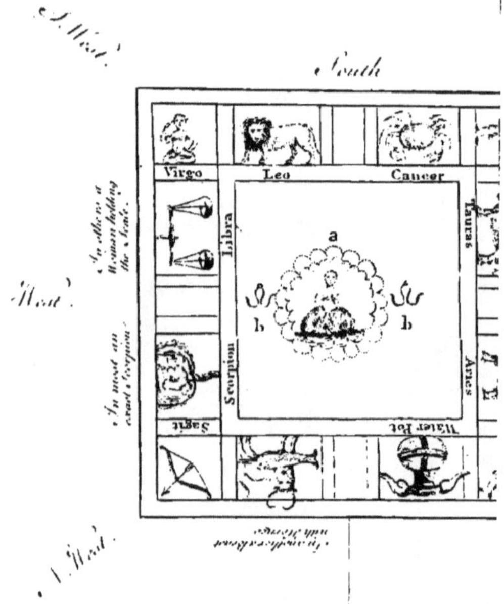

S. West.

South

Virgo

Leo

Cancer

Libra

Lauras

a

b

b

West.

In others a Woman holding the Scale.

In most an exact Scorpion.

Scorpion

Aries

Sagit

Water Pot

N. West.

a. *Symbol of the Universal*

b.b. *Two hooks of Iron to suspend a kind*

Deity or Swamy often sat, when exhibited

AN APPENDIX,

Containing an account of ſome experiments made by Mr. Hey, which prove that there is no oil of vitriol in water impregnated with fixed air extracted from chalk by oil of vitriol; and alſo a letter from Mr. Hey, to Dr. Prieſtley, concerning the effects of fixed air applied by way of clyſter.

EXPERIMENTS TO PROVE THAT THERE IS NO OIL OF VITRIOL IN WATER IMPREGNATED WITH FIXED AIR.

It having been ſuggeſted, that air ariſing from a fermenting mixture of chalk and oil of vitriol might carry up with it a ſmall portion of the vitriolic acid, rendered volatile by the act of fermentation; I made the following experiments, in order to diſcover whether the acidulous taſte, which water impregnated with ſuch air affords, was owing to the preſence of any acid, or only to the fixed air it had abſorbed.

EXPERIMENT I.

I mixed a tea-ſpoonful of ſyrup of violets with an ounce of diſtilled water, ſaturated with fixed air procured from chalk by means of the vitriolic acid; but neither upon the firſt mixture, nor after ſtanding 24 hours, was the colour of the ſyrup at all changed, except by its ſimple dilution.

Experiment II.

A portion of the ſame diſtilled water, unimpregnated with fixed air, was mixed with the ſyrup in the ſame proportion: not the leaſt difference in colour could be perceived betwixt this and the above mentioned mixture.

Experiment III.

One drop of oil of vitriol being mixed with a pint of the ſame diſtilled water, an ounce of this water was mixed with a tea-ſpoonful of the ſyrup. This mixture was very diſtinguiſhable in colour from the two former, having a purpliſh caſt, which the others wanted.

Experiment IV.

The diſtilled water impregnated with ſo ſmall a quantity of vitriolic acid having a more agreeable taſte than when alone, and yet manifeſting the preſence of an acid by means of the ſyrup of violets; I ſubjected it to ſome other teſts of acidity. It formed curds when agitated with ſoap, lathered with difficulty, and very imperfectly; but not the leaſt ebullition could be diſcovered upon dropping in ſpirit of ſal ammoniac, or ſolution of ſalt of tartar, though I had taken care to render the latter free from cauſticity by impregnating it with fixed air.

Ex-

EXPERIMENT V.

The diſtilled water ſaturated with fixed air neither effervefced, nor ſhewed any clouds, when mixed with the fixed or volatile alkali.

EXPERIMENT VI.

No curd was formed by pouring this water upon an equal quantity of milk, and boiling them together.

EXPERIMENT VII.

When agitated with ſoap, this water produced curds, and lathered with ſome difficulty; but not ſo much as the diſtilled water mixed with vitriolic acid in the very ſmall proportion above-mentioned. The ſame diſtilled water without any impregnation of fixed air lathered with ſoap without the leaſt previous curdling. River water, and a pleaſant pump water not remarkably hard, were compared with theſe. The former produced curds before it lathered, but not quite in ſo great a quantity as the diſtilled water impregnated with fixed air: the latter cauſed a ſtronger curd than any of the others above-mentioned.

EXPERIMENT VIII.

Apprehending that the fixed air in the diſtilled water occaſioned the coagulation, or ſeparation of the oily part of the ſoap, only by deſtroying the cauſticity of the *lixivium*, and thereby rendering the union lefs perfect betwixt that and the tallow, and not by the prefence of any acid; I impregnated a frefh parcel of the fame diftilled water with fixed air, which had paffed through half a yard of a wide barometer tube filled with falt of tartar; but this water caufed the fame curdling with foap as the former had done, and appeared in every refpect to be exactly the fame.

Experiment IX.

Diftilled water faturated with fixed air formed a white cloud and precipitation, upon being mixed with a folution of *faccharum faturni*. I found likewife, that fixed air, after paffing through the tube filled with alkaline falt, upon being let into a phial containing a folution of the metallic falt in diftilled water, caufed a perfect feparation of the lead, in form of a white powder; for the water, after this precipitation, fhewed no cloudinefs upon a frefh mixture of the fubftances which had before rendered it opaque.

A Letter

A Letter from Mr. Hey to Dr. Priestley, concerning the Effects of fixed Air applied by way of Clyster.

Leeds, Feb. 15th, 1772.

Reverend Sir,

Having lately experienced the good effects of fixed air in a putrid fever, applied in a manner, I believe, not heretofore made use of, I thought it proper to inform you of the agreeable event, as the method of applying this powerful corrector of putrefaction took its rise principally from your observations and experiments on factitious air; and now, at your request, I send the particulars of the case I mentioned to you, as far as concerns the administration of this remedy.

January 8, 1772, Mr. Lightbowne, a young gentleman who lives with me, was seized with a fever, which, after continuing about ten days, began to be attended with those symptoms that indicate a putrescent state of the fluids.

18th, His tongue was black in the morning when I first visited him, but the blackness went off in the day-time upon drinking: He had begun to doze much the preceding day, and now he took little notice of those that were about him: His belly was loose, and had been so for some days: his pulse beat 110 strokes in a minute, and was rather low: he was ordered to take twenty five grains of Peruvian bark with five of tormentill root in powder every four hours, and to use red wine and water cold as his common drink.

19th, I was called to viſit him early in the morning, on account of a bleeding at the noſe which had come on: he loſt about eight ounces of blood, which was of a looſe texture: the hæmorrhage was ſuppreſſed, though not without ſome difficulty, by means of tents made of ſoft lint, dipped in cold water ſtrongly impregnated with tincture of iron, which were introduced within the noſtrils quite through to their poſterior apertures; a method which has never yet failed me in like caſes. His tongue was now covered with a thick black pellicle, which was not diminiſhed by drinking: his teeth were furred with the ſame kind of ſordid matter, and even the roof of his mouth and fauces were not free from it: his looſeneſs and ſtupor continued, and he was almoſt inceſſantly muttering to himſelf: he took this day a ſcruple of the Peruvian bark with ten grains of tormentill every two or three hours: a ſtarch clyſter containing a drachm of the compound powder of bole, without opium, was given morning and evening: a window was ſet open in his room, though it was a ſevere froſt, and the floor was frequently ſprinkled with vinegar.

20th, He continued nearly in the ſame ſtate: when rouzed from his dozing, he generally gave a ſenſible anſwer to the queſtions aſked him; but he immediately relapſed, and repeated his muttering. His ſkin was dry, and harſh, but without *petechiæ*. He ſometimes voided his urine and *fæces* into the bed, but generally had ſenſe enough to aſk for the bed-pan: as he now nauſeated the bark in ſubſtance, it was exchanged for Huxham's

tincture,

tincture, of which he took a table-ſpoonful every two hours in a cup full of cold water: he drank ſometimes a little of the tincture of roſes, but his common liquors were red wine and water, or rice water and brandy acidulated with elixir of vitriol: before drinking, he was commonly requeſted to rinſe his mouth with water to which a little honey and vinegar had been added. His looſeneſs rather increaſed, and the ſtools were watery, black, and fœtid: It was judged neceſſary to moderate this diſcharge, which ſeemed to ſink him, by mixing a drachm of the *theriaca Andromachi* with each clyſter.

21ſt. The ſame putrid ſymptoms remained, and a *ſubſultus tendinum* came on: his ſtools were more fœtid; and ſo hot, that the nurſe aſſured me ſhe could not apply her hand to the bed-pan, immediately after they were diſcharged, without feeling pain on this account: The medicine and clyſters were repeated.

Reflecting upon the diſagreeable neceſſity we ſeemed to lie under of confining this putrid matter in the inteſtines, leſt the evacuation ſhould deſtroy the *vis vitæ* before there was time to correct its bad quality, and overcome its bad effects, by the means we were uſing; I conſidered, that, if this putrid ferment could be more immediately corrected, a ſtop would probably be put to the flux, which ſeemed to ariſe from, or at leaſt to be encreaſed by it; and the *fomes* of the diſeaſe would likewiſe be in a great meaſure removed. I thought nothing was ſo likely to effect this, as the introduction of fixed air into the alimentary canal,

 which,

which, from the experiments of Dr. Macbride, and thofe you have made fince his publication, appears to be the moft powerful corrector of putrefaction hitherto known. I recollected what you had recommended to me as deferving to be tried in putrid difeafes, I mean, the injection of this kind of air by way of clyfter, and judged that in the prefent cafe fuch a method was clearly indicated.

The next morning I mentioned my reflections to Dr. Hird and Dr. Crowther, who kindly attended this young gentleman at my requeft, and propofed the following method of treatment, which, with their approbation, was immediately entered upon. We firft gave him five grains of ipecacoanha, to evacuate in the moft eafy manner part of the putrid *colluvies*: he was then allowed to drink freely of brifk orange-wine, which contained a good deal of fixed air, yet had not loft its fweetnefs: the tincture of bark was continued as before; and the water, which he drank along with it, was impregnated with fixed air from the atmofphere of a large vat of fermenting wort, in the manner I had learned from you: inftead of the aftringent, air alone was injected, collected from a fermenting mixture of chalk and oil of vitriol: he drank a bottle of orange-wine in the courfe of this day, but refufed any other liquor except water and his medicine: two bladders full of air were thrown up in the afternoon.

23d. His ftools were lefs frequent; their heat likewife and peculiar *fœtor* were confiderably diminifhed: his muttering was much abated, and the *fubfultus tendinum* had left him. Finding that part of the air was rejected when given with a bladder in the

the uſual way, I contrived a method of injecting it which was not ſo liable to this inconvenience. I took the flexible tube of that inſtrument which is uſed for throwing up the fume of tobacco, and tied a ſmall bladder to the end of it that is connected with the box made for receiving the tobacco, which I had previouſly taken off from the tube: I then put ſome bits of chalk into a ſix ounce phial until it was half filled; upon theſe I poured ſuch a quantity of oil of vitriol as I thought capable of ſaturating the chalk, and immediately tied the bladder, which I had fixed to the tube, round the neck of the phial: the clyſter pipe, which was faſtened to the other end of the tube, was introduced into the *anus* before the oil of vitriol was poured upon the chalk. By this method the air paſſed gradually into the inteſtines as it was generated; the rejection of it was in a great meaſure prevented; and the inconvenience of keeping the patient uncovered during the operation was avoided.

24th, He was ſo much better, that there ſeemed to be no neceſſity for repeating the clyſters: the other means were continued. The window of his room was now kept ſhut.

25th, All the ſymptoms of putreſcency had left him; his tongue and teeth were clean; there remained no unnatural blackneſs or *fœtor* in his ſtools, which had now regained their proper conſiſtence; his dozing and muttering were gone off; and the diſagreeable odour of his breath and perſpiration was no longer perceived. He took nouriſhment to-day, with pleaſure; and, in the afternoon, ſat up an hour in his chair.

His

His fever, however, did not immediately leave him; but this we attributed to his having caught cold from being incautiouſly uncovered, when the window was open, and the weather extremely ſevere; for a cough, which had troubled him in ſome degree from the beginning, increaſed, and he became likewiſe very hoarſe for ſeveral days, his pulſe, at the ſame time, growing quicker: but theſe complaints alſo went off, and he recovered, without any return of the bad ſymtoms above-mentioned.

I am, Reverend Sir,

Your obliged humble ſervant,

Wm Hey.

P. S.

October 29, 1772.

Fevers of the putrid kind have been ſo rare in this town, and in its neighbourhood, ſince the commencement of the preſent year, that I have not had an opportunity of trying again the effects of fixed air, given by way of clyſter, in any caſe exactly ſimilar to Mr. Lightbowne's. I have twice given water ſaturated with fixed air in a fever of the putreſcent kind, and it agreed very well with the patients. To one of them the aërial clyſters were adminiſtred, on account of a looſeneſs, which attended the fever, though the ſtools were not black, nor remarkably hot or fœtid.

These

These clysters did not remove the looseness, though there was often a greater interval than usual betwixt the evacuations, after the injection of them. The patient never complained of any uneasy distention of the belly from the air thrown up, which, indeed, is not to be wondered at, considering how readily this kind of air is absorbed by aqueous and other fluids, for which sufficient time was given, by the gradual manner of injecting it. Both those patients recovered, though the use of fixed air did not produce a crisis before the period on which such fevers usually terminate. They had neither of them the opportunity of drinking such wine as Mr. Lightbowne took after the use of fixed air was entered upon; and this, probably, was some disadvantage to them.

I find the methods of procuring fixed air, and impregnating water with it, which you have published, are preferable to those I made use of in Mr. Lightbowne's case.

The flexible tube used for conveying the fume of tobacco into the intestines, I find to be a very convenient instrument in this case, by the method before-mentioned (only adding water to the chalk, before the oil of vitriol is instilled, as you direct): the injection of air may be continued at pleasure, without any other inconvenience to the patient, than what may arise from his continuing in one position during the operation, which scarcely deserves to be mentioned, or from the continuance of the clyster-pipe within the anus, which is but trifling, if it be not shaken much, or pushed against the rectum.

When I said in my letter, that fixed air appeared to be the greatest corrector of putrefaction hitherto known, your philoſophical reſearches had not then made you acquainted with that moſt remarkably antiſeptic property of nitrous air. Since you favoured me with a view of ſome aſtoniſhing proofs of this, I have conceived hopes, that this kind of air may likewiſe be applied medicinally to great advantage.

W. H.

A Correction.

Upon re-examining Dr. Hales's account of his experiments to meaſure the diminution of air by reſpiration (Statical Eſſays, Vol. I. p. 238, 4th edition), I find an error of the preſs, of $\frac{3}{13}$ for $\frac{1}{13}$; ſo that the diminution of air by reſpiration, though very various, is, I believe, always conſiderably *leſs* than by putrefaction, or ſeveral other cauſes of diminution. But though I have mentioned this diminution as equal to ſeveral others, nothing material depends upon it; the quality of the air thus diminiſhed being, in all reſpects, the ſame, notwithſtanding the cauſe of increaſe (which, as I have obſerved, in this and other caſes, co-operates with the cauſe of diminution) be greater than I had ſuppoſed.

I did not endeavour to meaſure the quantity of the diminution of air by reſpiration, as I did that by other cauſes; becauſe I imagined that it had been done ſufficiently by others, and eſpecially by Dr. Hales.

Received November 29, 1771.

XX. *An Essay on the periodical Appearing and Disappearing of certain Birds, at different Times of the Year. In a Letter from the Honourable* Daines Barrington, *Vice-Pres. R. S. to* William Watson, *M. D. F. R. S.*

DEAR SIR,

Read April 2, 9, 30, and May 14, 1772.

AS I know, from some conversation we have had on this head, that you consider the migration of birds as a very interesting point in natural history, I send you the following reflections on this subject as they have occurred to me upon looking into most of the ornithologists who have written on this question.

It will be first necessary in the present, as in all other disputes, to define the terms on which the controversy arises. I therefore premise that I mean by the word Migration, a periodical passage by a whole species of birds across a considerable extent of sea.

I do not mean therefore to deny that a bird, or birds, may possibly fly now and then from Dover to Calais,

Calais, from Gibraltar to Tangier, or any other ſuch narrow ſtrait, as the oppoſite coaſts are clearly within the bird's ken, and the paſſage is no more adventurous than acroſs a large freſh water lake.

I as little mean to deny that there may be a periodical flitting of certain birds from one part of a continent to another: the Royſton Crow, and Rock Ouzel, furniſh inſtances of ſuch a regular migration.

What I mean chiefly to contend therefore is, that it ſeems to be highly improbable, birds ſhould, at certain ſeaſons, traverſe large tracts of ſea, or rather ocean, without leaving any of the ſame ſpecies behind, but the ſick or wounded.

As this litigated point can only receive a ſatisfactory deciſion from very accurate obſervations, all preceding naturaliſts, from Ariſtotle to Ray, have ſpoken with much doubt concerning it.

Soon after the appearance of Monſ. Adanſon's voyage to Senegal, however, Mr. Collinſon firſt, in the Philoſophical Tranſactions *, and after him the moſt eminent ornithologiſts of Europe, ſeem to have conſidered this traveller's having caught four European Swallows on the 6th of October, not far from the African coaſt, as a deciſive proof, that the common ſwallows, when they diſappear in Europe, make for Africa during the winter, and return again to us in the ſpring.

It is therefore highly incumbent upon me, who profeſs that I am by no means ſatisfied with the account, given by Monſ. Adanſon of theſe European

* Part II. 1760, p. 459, & ſeq.

ſwallows,

ſwallows, to enter into a very minute diſcuſſion of what may, or may not, be inferred from his obſervation according to his own narrative.

I ſhall firſt however conſider the general arguments, from which it is ſuppoſed that birds of paſſage periodically traverſe oceans, which indeed may be almoſt reduced to this ſingle one, *viz.* we ſee certain birds in particular ſeaſons, and afterwards we ſee them not; from which data it is at once inferred, that the cauſe of their diſappearance is, that they have croſſed large tracts of ſea.

The obvious anſwer to this is, that no well-atteſted inſtances can be produced of ſuch a migration, as I ſhall endeavour to ſhew hereafter; but beſides this convincing negative proof, there are not others wanting.

Thoſe who ſend birds periodically acroſs the ſea, being preſſed with the very obvious anſwer I have before ſuggeſted, have recourſe to two ſuppoſitions, by which they would account for their not being obſerved by ſeamen during their paſſage.

The firſt is, that they riſe ſo high in the air that they become inviſible *; but unfortunately the riſing to this extraordinary height, or the falling from it, is equally deſtitute of any ocular proof, as the birds being ſeen during their paſſage.

I have indeed converſed with ſome people, who conceive they have loſt ſight of birds by their perpendicular flight; I muſt own, however, that I have

* It is well known that ſome ornithologiſts have even ſuppoſed that they leave our atmoſphere for that of the Moon. See Harl. Miſc. Vol. II. p. 561.

always ſuppoſed them to be ſhort-ſighted, as I never loſt the ſight of a bird myſelf, but from its horizontal diſtance, and I doubt much whether any bird was ever ſeen to riſe to a greater height than perhaps twice that of St. Paul's croſs *.

There ſeems to be but one method indeed, by which the height of a bird in the air may be eſtimated; which is, by comparing its apparent ſize with its known one, when very near us; and it need not be ſaid that method of calculating muſt depend entirely upon the ſight of the obſerver, who, if he happens not to ſee objects well at a diſtance, will very ſoon ſuppoſe the bird to be loſt in the clouds.

There is alſo another objection to the hypotheſis of birds paſſing ſeas at ſuch an extraordinary height, ariſing from the known rarefaction of the air, which may poſſibly be inconvenient for reſpiration, as well as flight; and if this was not really the caſe, one ſhould ſuppoſe that birds would frequently riſe to ſuch uncommon elevations, when they had no occaſion to traverſe oceans.

* Wild geeſe fly at the greateſt height of any bird I ever happened to attend to; and from comparing them with rooks, which I have frequently looked at, when perched on the croſs of St. Paul's, I cannot think that a wild-gooſe was ever diminiſhed, to my ſight at leaſt, more than he would be at twice the height of St. Paul's, or perhaps 300 yards. Mr. Hunter, F. R. S. informs me, that the bird which hath appeared to him as the higheſt flier, is a ſmall eagle on the confines of Spain and Portugal, which frequents high rocks. Mr. Hunter hath firſt ſeen this ſpecies of eagle from the bottom of a mountain, and followed it to the top, when the bird hath riſen ſo high as to appear leſs than he did from the bottom. Mr. Hunter however adds, that he could ſtill hear the cry, and diſtinguiſh the bird.

The

The Scotch Ptarmigan frequents the higheſt ground of any Britiſh bird, and he takes but very ſhort flights.

But it is alſo urged by ſome, that the reaſon why ſeamen do not regularly ſee the migration of birds, is becauſe they chooſe the night, and not the day, for the paſſage *.

Now though it may be allowed, that poſſibly birds may croſs from the coaſt of Holland to the Eaſtern coaſt of England (for example) during a long night, yet it muſt be dark nearly as long as it is within the Arctic circle to afford time for a bird to paſs from the Line to many parts of Europe, which Monſ. de Buffon calculates, may be done in about eight or nine days †.

If the paſſage happened in half the nights of the year, which have the benefit of moonlight, the birds would be diſcovered by the ſailors almoſt as well as in the day time; to which I muſt add that ſeveral ſuppoſed birds of paſſage (the Fieldfare in particular) always call when on their flight, ſo that the ſeamen muſt be deaf as well as blind, if ſuch flocks of birds eſcape their notice.

Other objections however remain to this hypotheſis of a paſſage during the night.

* Mr. Cateſby ſuppoſes that they may thus paſs in the night time, to avoid birds of prey. Phil. Tranſ. Abr. Vol. II. p. 887. But are not owls then ſtirring?

On the other hand, if they migrate in the day time, kites, hawks, and other birds of prey, muſt be very bad ſportſmen not to attend (like Arabs) theſe large and periodical caravans.

† In the preface to the firſt volume of his lately publiſhed Ornithology, p. 32.

Ninety-

Moſt birds not only ſleep during the night, but are as much incapacitated from diſtinguiſhing objects well as we are, in the abſence of the ſun: it is therefore inconceivable that they ſhould chooſe owl-light for ſuch a diſtant journey.

Beſides this, the Eaſtern coaſt of England, to which birds of paſſage muſt neceſſarily firſt come from the continent, hath many light-houſes upon it; they would therefore, in a dark night, immediately make for ſuch an object, and deſtroy themſelves by flying with violence againſt it, as is well known to every bat-fowler.

Having endeavoured to anſwer theſe two ſuppoſitions, by which it is contended that birds of paſſage may eſcape obſervation in their flight; I ſhall now conſider all the inſtances I have been able to meet with of any birds being actually ſeen whilſt they were croſſing any extent of ſea, though I might give a very ſhort refutation to them, by inſiſting, that if this was ever experienced, it muſt happen as conſtantly in a ſea, which is much navigated, as the return of the ſeaſons.

I cannot do better than to follow theſe according to chronological order.

The firſt in point of time is that which is cited by Willoughby *, from Bellon, whoſe words are thus tranſlated, " When we ſailed from Rhodes to " Alexandria, many quails flying from the North " towards the South, were taken in our ſhip, whence " I am perſuaded that they ſhift places; for for- " merly, when I ſailed out of the Iſle of Zant to " Morea, or Negropont, in the ſpring, I had ob-

* B. II. c. 11. §. 8.

" ſerved

" ſerved quails flying the contrary way to N. and S. " that they might abide there all ſummer, at which " time alſo a great many were taken in the ſhip."

Let us now conſider what is to be inferred from this citation.

In the firſt place, Bellon does not particularize the longitude and latitude of that part of the Mediterranean, which he was then croſſing; and in his courſe from Rhodes to Alexandria, both the iſlands of Scarpanto and Crete could be at no great diſtance: theſe quails therefore were probably flitting from one iſland of the Mediteranean * to another.

The ſame obſervation may be made with regard to the quails which he ſaw between Zant and Negropont, as the whole paſſage is crouded with iſlands, they therefore might be paſſing from iſland to iſland, or headland to headland, which might very probably lye Eaſt and Weſt, ſo as to occaſion the birds flying in a different direction, from which they paſſed the ſhip before.

I have therefore no objection to this proof of migration, if it is only inſiſted upon to ſhew that a quail ſhifts its ſtation at certain ſeaſons of the year; but cannot admit that it is fair from hence to argue that theſe birds periodically croſs large tracts of ſea.

Bellon himſelf ſtates, that when the birds ſettled upon the ſhip, they were taken by the firſt perſon who choſe to catch them, and therefore they muſt have been unequal to the ſhort flight which they were attempting.

* One of the Mediterranean iſlands is ſuppoſed to have obtained its ancient name of Ortygia from the numbers of quails.

It

It is very true that quails have been often pitched upon as inftances of birds that migrate acrofs feas, becaufe they are fcarcely ever feen in winter: it is well known, however, to every fportsman, that this bird never flies 300 yards at a time, and the tail being fo fhort, it is highly improbable they fhould be equal to a paffage of any length.

We find therefore, that quails, which are commonly fuppofed to leave our ifland in the winter, in reality retire to the fea coafts, and pick up their food amongft the fea weeds *.

I have happened lately to fee a fpecimen of a particular fpecies of quail, which is defcribed by Dr. Shaw †, and is diftinguifhed from the other kinds by wanting the hind-claw.

Dr. Shaw alfo ftates that it is a bird of paffage. Now if quails really migrate from the coaft of Barbary to Italy, as is commonly fuppofed, whence can it have arifen that this remarkable fpecies hath efcaped the notice of Aldrovandus, Olina, and the other Italian ornithologifts?

When I had juft finifhed what I have here faid with regard to the migration of quails, I have had an opportunity of feeing the fecond volume of Monf. de Buffon's ornithology ‡; where, under this article, he contends that this bird leaves Europe in the winter.

It is incumbent upon me, therefore, either to own I am convinced by what this moft ingenious and able naturalift hath urged, or to give my reafons why I

* See Br. Zool. Vol. II. p. 210. 2d Ed. octavo.

† Phyf. Obf. on the kingdom of Algiers, ch. 2.

‡ See p. 459, & feq.

ftill

ſtill continue to diſſent from the opinion he maintains.

Though M. de Buffon hath diſcuſſed this point very much at large, yet I find only the following facts or arguments to be new.

He firſt cites the Memoirs of the Academy of Sciences *, for an account given by M. Godeheu of quails coming to the iſland of Malta in the month of May, and leaving it in September.

The firſt anſwer to this obſervation is, that the iſland of Malta is not only near to the coaſt of Africa, but to ſeveral of the Mediterranean iſlands; it therefore amounts to no more than the flitting I have before taken notice of †.

Monſ. de Buffon ſuppoſes that a quail only quits one latitude for another, in order to meet with a perpetual crop on the ground.

Now can it be ſuppoſed that there is that difference between the harveſt on the coaſt of Africa, and that of the ſmall quantity of grain which grows on the rocky iſland of Malta, that it becomes inconvenient to the bird to ſtay in Africa as ſoon as May ſets in; and neceſſary, on the other hand, to continue in Malta from May till September.

Monſ. de Buffon then ſuppoſes that quails make their paſſage in the night, as well as conceives them to be of a remarkably warm temperature ‡, and ſays

* Tom. III. p. 91 and 92.

† Both Monſ. de Godeheu and M. de Buffon ſeem to conceive that the quail ſhould fly in the ſame direction as the wind blows; but birds on the wing from point to point, which are at a conſiderable diſtance, fly againſt the wind, as their plumage is otherwiſe ruffled.

‡ As this is given for a reaſon why the African quails migrate Northward: Q. what is to become of the Icelandic quails during the ſummer?

 that

that " *chaud comme une caille*," is in every one's mouth *.

Now in the first place their migration during the night, is contrary to Belon's account, which M. de Buffon so much relies upon, who expresly says, that the birds were caught in the day time †.

In the next place, I apprehend that " *chaud comme* " *une caille*," alludes to the very remarkable salaciousness of this bird, and not to the constant heat of its body.

Monf. de Buffon then observes, that if quails are kept in a cage, they are remarkably impatient of confinement in the autumn and spring, whence he infers that they then want to migrate ‡; he also adds, in the same period, that this uneasiness begins an hour before the sun rises, and that it continues all the night.

This great naturalist does not state this observation as having been made by himself, and it seems upon the face of it to be a very extraordinary one.

* All birds indeed are warmer by four degrees than other animals. See some ingenious thermometrical experiments by Mr. Martin of Aberdeen, Edinb. 1771, 12mo.

† Upon looking a second time into Belon, he does not indeed state whether it was in the day or the night; but if it had happened in the latter, this traveller and ornithologist could not well have omitted such a circumstance. Besides this, he mentions in what direction the quails were flying, which he could not have discerned in the night.

‡ It may also arise from this bird's being of so quarrelsome a disposition, and consequently most likely to fight with its fellow prisoners when they are all in greatest vigour after moulting, and on the return of the spring.

M. de Buffon allows that they will fight for a grain of millet, and adds, " car parmi les animaux il faut un sujet reel pour se " battre." M. de Buffon hath never been in a cockpit.

No one (at leaſt with us) ever keeps quails in a cage except the poulterers, who always ſell them as faſt as they are fat, and conſequently can give no account of what happens to them during ſo long an impriſonment as this obſervation neceſſarily implies.

No ſuch remarkable uneaſineſs hath ever been attended to in any other ſuppoſed bird of paſſage during its confinement; but, allowing the fact to be as M. de Buffon ſtates, he himſelf ſupplies us with the real cauſe of this impatience.

He aſſerts, that quails conſtantly moult twice * a year, *viz.* at the cloſe both of ſummer and winter; whence it follows, that the bird, in autumn and the ſpring, muſt be in full vigour upon its recovery from this periodical illneſs: it can therefore as little brook confinement, as the phyſician's patient upon the return of health after illneſs.

Thus much I have thought it neceſſary to ſay, in anſwer to M. de Buffon, who "dum errat, docet," who ſcarcely ever argues ill but when he is miſinformed as to facts, and who often, from ſtrength of underſtanding, diſbelieves ſuch intelligence as might impoſe upon a naturaliſt of leſs acuteneſs and penetration.

* I have often heard that certain birds moult twice a year, ſome of which I have kept myſelf without their changing their feathers more than once.

I ſhould ſuppoſe that this notion ariſes from ſome birds not moulting regularly in the autumn every year; and when the change takes place in the following ſpring, they very commonly die: I can ſcarcely think that many of them are equal to two illneſſes of ſo long a continuance, which are conſtantly to return within twelvemonths.

I ſhould therefore rather account for the extraordinary briſkneſs of a quail in autumn and the ſpring, from its recovery after moulting in the former, and from the known effects of the ſpring as to moſt animals in the latter.

 The

The next inſtance of a bird being caught at any diſtance from land, is in Sir Hans Sloane's voyage to Jamaica, who ſays, that a lark was taken in the ſhip 40 leagues from the ſhore: this therefore was certainly an unfortunate bird, forced out to ſea by a ſtrong wind in flying from headland to headland, as no one ſuppoſes the ſkylark to be a bird of paſſage.

The ſame anſwer may be given to a yellow-hammer's ſettling upon Haſſelquiſt's ſhip in the entrance of the Mediterranean, with this difference, that either the European or African coaſt muſt have been much nearer than 40 leagues *.

The next fact to be conſidered is what is mentioned in a letter of Mr. Peter Collinſon's, printed in the Philoſophical Tranſactions †.

He there ſays, " That Sir Charles Wager had " frequently informed him, that in one of his " voyages home in the ſpring as he came into ſound- " ings in our channel, that a great flock of ſwallows " almoſt covered his rigging, that they were nearly " ſpent and famiſhed, and were only feathers and " bones; but being recruited by a night's reſt, they " took their flight in the morning."

The firſt anſwer to this is, that if theſe were birds which had croſſed large tracts of ſea in their periodical migrations, the ſame accident muſt happen eternally, both in the ſpring and autumn, which is not however pretended by any one.

In the next place, the ſwallows are ſtated to be ſpent both by famine and fatigue; and how were they to procure any flies or other ſuſtenance on the

* See Haſſelquiſt's Travels, in princ.

† 1760, Part II. p. 461.

rigging of the admiral's ſhip, though they migth indeed reſt themſelves?

Sir Charles, however, expreſly informs us, that he was in the channel, and within ſoundings: theſe birds, therefore (like Bellon's quails) were only paſſing probably from headland to headland; and being forced out by a ſtrong wind, were obliged to ſettle upon the firſt ſhip they ſaw, or otherwiſe muſt have dropped into the ſea, which I make no doubt happens to many unfortunate birds under the ſame circumſtances.

As the birds which thus ſettled upon Sir Charles Wager's rigging were ſwallows, it very naturally brings me now to conſider the celebrated obſervation of Monſ. Adanſon, under all its circumſtances, as it hath been ſo much relied upon, and by naturaliſts of ſo great eminence.

Monſ. Adanſon is a very ingenious writer, and the publick is much indebted to him for many of the remarks which he made whilſt he reſided in Senegal.

I may, however, I think, preſume to ſay, that he had not before his voyage made ornithology his particular ſtudy; proofs of which are not wanting in other parts of his work, which do not relate to ſwallows.

For example, he ſuppoſes, that the Canary birds which are bred in Europe are white, and that they become ſo by our climate's being more cold than that of Africa.

"J'ai remarqué que le ſerin qui devient tout blanc "en France, eſt a Teneriffe d'un gris preſque auſſi "foncé que celui de la linotte; ce changement de "couleur provient vraiſemblablement de la froidure "de notre climat *."

* Voyage au Senegal, p. 13.

Mr.

Mr. Adanſon in this paſſage ſeems to have deduced two falſe inferences from having ſeen a few white Canary birds in France, which he afterwards compares with thoſe of Teneriff, and ſuppoſes the change of colour to ariſe merely from alteration of climate: it is known, however, almoſt to every one, that there is an infinite variety in the plumage of the European Canary birds, which, as in poultry, ariſes from their being pampered with ſo much food, as well as confinement *.

Monſ. Adanſon, in another part of his voyage †, deſcribes a Roller, which he ſuppoſes to migrate ſometimes to the Southern parts of Europe.

This circumſtance ſhews that he could not have looked much into books of natural hiſtory, becauſe the principal ſynonym of this bird is *garrulus Argentoratenſis* ‡; and Linnæus informs us that it is found even in Sweden ||.

* In the ſame paſſage, he compares the colour of the African Canary bird to that of the European linnet, and ſays it is *d'un gris preſque auſſi foncé*, whereas the European linnet is well known to be brown, and not grey. The linnet affords a very deciſive proof that the change of plumage does not ariſe from the difference of climate, but the two cauſes I have aſſigned. The cock bird, whilſt at liberty, hath a red breaſt: yet if it is either bred up in a cage from the neſt, or is caught with its red plumage, and afterwards moults in the houſe, it never recovers the red feathers.

That moſt able naturaliſt, Monſ. de Buffon, from having ſeen ſome cock linnets which had thus moulted off, or perhaps ſome hen linnets (which have not a red breaſt) conſiders them as a diſtinct ſpecies, and compares their breeding together in an aviary, to that of the Canary bird and goldfinch. Ornith. p. xxii.

† P. 16. ‡ Or of Straſburgh.

|| Faun. Suec. 94.

The

The ſtrong characteriſtic mark of this bird, is the outermoſt feathers of the tail, which able naturaliſts deſcribe as three fourths of an inch longer than the reſt *. Monſ. Adanſon, however, compares their length, not with the other feathers of the tail, but with the length of the bird's body, which is by no means the natural or proper ſtandard of compariſon.

The reaſon of my taking notice of theſe more minute inaccuracies in Monſ. Adanſon's account of birds, ariſes from Mr. Collinſon's relying upon his obſervations with regard to ſwallows being ſo abſolutely deciſive, becauſe he is repreſented to be ſo able a naturaliſt.

I ſhall now ſtate (very minutely) under what circumſtances theſe ſwallows were caught, and what ſeems to be the true inference from his own account.

He informs us, that four ſwallows ſettled upon the ſhip, not 50 leagues from the coaſt of Senegal, on the 6th of October; that theſe birds were taken, and that he knew them to be the true ſwallow of Europe †, which he ſuppoſes were then returning to the coaſt of Africa.

I ſhall now endeavour to ſhew that theſe birds could not be European ſwallows; nor, if they were, could they have been on their return from Europe to Africa.

* Willoughby, p. 131. Br. Zool. Vol. II. in append.

† I have before endeavoured to ſhew that Monſ. Adanſon does not always recollect with accuracy the plumage of the moſt common European birds, by what he ſays with regard to the linnet.

The

The word *hirondelle*, in French, is uſed as a general term for the four * ſpecies of theſe birds, as the term *ſwallow* is with us.

Now the four ſwallows thus caught and examined by Monſ. Adanſon were either all of the ſame ſpecies, or intermixed in ſome other proportion.

Would not then any naturaliſt in ſtating ſo material a fact (as he himſelf ſuppoſes it to be) have particularized of what ſpecies of ſwallow theſe very intereſting birds were?

Should not Monſ. Adanſon alſo have taken care to diſtinguiſh theſe ſuppoſed European ſwallows from two ſpecies of the ſame tribe, which bear a general reſemblance to thoſe of Europe, and are not only deſcribed, but engraved by Briſſon, under the name of *Hirondelle de Senegal & Hirondelle de rivage du Senegal* † ?

Though Monſ. Adanſon was above a year on this part of the African coaſt, paid ſo much attention to ſwallows, and was ſo immediately acquainted with the different ſpecies on the firſt inſpection, yet he ſeems never to have diſcovered that there were ſuch African ſwallows as are thus deſcribed and engraved by Briſſon, though he muſt have ſeen them daily.

Monſ. Adanſon however concludes his account of the ſuppoſed European ſwallow, whilſt it continues on the coaſt of Senegal, by a circumſtance which

* *Viz.* the ſwallow κατ' ἐξοχήν, the martin, the ſand martin, and the ſwift: I omit the goatſucker, becauſe this bird, though properly claſſed as a ſpecies of ſwallow by ornithologiſts, is not ſo conſidered by others.

† See Briſſon, Tom. II. pl. xiv.

ſeems

ſeems to prove to demonſtration of what ſpecies the four ſwallows caught in the ſhip really were.

He ſays that they rooſt on the ſand either by themſelves, or at moſt only in pairs, and that they frequent the coaſt much more than the inland parts *.

Theſe ſwallows, therefore, if they came from Europe, muſt have immediately changed at once their known habits: and is it not conſequently moſt clear that they were of that ſpecies which Briſſon deſcribes under the name of *Hirondelle de rivage du Senegal*?

But though it ſhould be admitted, notwithſtanding what I have inſiſted upon, from Monſ. Adanſon's own account, that theſe were really ſwallows of the ſame kind with thoſe of Europe; yet I muſt ſtill contend that they could not poſſibly have been on their return from Europe to Africa, becauſe the high road for a bird from the moſt Weſtern point of Europe to Senegal, is along the N. Weſt coaſt of Africa, which projects greatly to the Weſtward of any part of Europe.

What then could be the inducement to theſe four ſwallows to fly 50 leagues to the Weſtward of the coaſt of Senegal, ſo much out of the proper direction?

It ſeems to me therefore, very clear, that theſe ſwallows (whether of the European kind or not) were flitting from the cape de Verde iſlands to the

* Voyage au Senegal, p. 67. I wiſh Monſ. Adanſon had alſo informed us whether theſe ſwallows had the ſame notes with thoſe of Europe, which is a very material circumſtance in the natural hiſtory of birds, though little attended to by moſt ornithologiſts.

 coaſt

coaſt of Africa, to which ſhort flight, however, they were unequal, and were obliged from fatigue to fall into the ſailors hands.

Monſ. Adanſon likewiſe mentions * that the ſhip's company caught a Roller on the 26th of April, which he ſuppoſes was on its paſſage to Europe, though he was then within ſight of the coaſt of Senegal: this bird, however, muſt be admitted not to have had ſufficient ſtrength to reach the firſt ſtage of this round-about journey, and was therefore probably forced out to ſea by a ſtrong wind, in paſſing from head-land to head-land.

But I muſt not diſmiſs what hath been obſerved with regard to the ſwallows ſeen by Monſ. Adanſon at Senegal, without endeavouring alſo to anſwer what M. de Buffon hath not only inferred from it, but hath endeavoured to confirm by an actual experiment †.

M. de Buffon, from the many inſtances of ſwallows being found torpid even under water, very readily admits, that all the birds of this genus do not migrate, but only that ſpecies which was ſeen by Monſ. Adanſon in Africa, and which he generally refers to as the chimney ſwallow ‡; but from the outſet, ſeems

* Voyage au Senegal, p. 15.

† See the two prefatory diſcourſes to his ſixteenth volume of natural hiſtory.

‡ So little do naturaliſts know of this very common bird, that I believe it hath never yet been obſerved by any writer, that the male ſwallow hath only the long ſlender feathers in the tail, which are conſidered as its moſt diſtinguiſhing marks. I venture to make this remark upon having ſeen the difference in two ſwallows which are in Mr. Tunſtall's collection, F. R. S. as alſo in two others, which have lately been preſented to the Muſeum

to ſhew that he hath himſelf confounded this ſpecies with the martin.

" Prenons un ſeul oiſeau, par exemple, l'hiron-
" delle, celle que tout le monde connoit, qui paroit
" au printems, diſparoit en automne, & fait ſon nid
" avec de la terre contre les fenetres, ou dans les
" cheminees." p. 23.

It is very clear that the deſign in this period is to ſpecify a particular bird in ſuch a manner that no doubt could remain with any one about the ſpecies referred to; and from other paſſages which follow, it is as clear that Monſ. de Buffon means to allude to the ſwallow κατ' εξοχην.

Though this was certainly the intention of this moſt ingenious naturaliſt, it is to me very evident that the martin, and not the ſwallow, was in his contemplation, becauſe he firſt ſpeaks of the bird's building againſt windows, before he mentions chimneys, and therefore ſuppoſes that either place is indifferent; which is not the caſe, becauſe the ſwallow ſeldom builds on the ſides of windows, or the martin in chimneys.

There are perhaps three or four martins to one ſwallow in all parts; and from their being the more common bird of the two, as well as from the circumſtance of their building at the corner of windows (and conſequently being eternally in our ſight), nine-

of the Royal Society, by the directors of the Hudſon's Bay company.

These long feathers would be very inconvenient to the hen during incubation; and they are likewiſe confined to the cock *widow-bird*, as, from their more extraordinary length, they would be ſtill more ſo.

teen out of twenty, when they ſpeak of a ſwallow, really mean a martin *.

I only take notice of this ſuppoſed inacuracy in Monſ. de Buffon, becauſe, if that able naturaliſt does not ſpeak of the different ſorts of ſwallows with that preciſion which is neceſſary upon ſuch an occaſion, why ſhould he rely ſo intirely upon the impoſſibility of Monſ. Adanſon's being miſtaken?

I ſhall now ſtate the experiment of Monſ. de Buffon, to prove that the ſwallow is not torpid in the winter, and muſt therefore migrate to the coaſt of Senegal †.

He ſhut up ſome ſwallows *(hirondelles)* in an ice houſe, which were there confined "plus ou moins "de temps;" and the conſequence was, that thoſe which remained there the longeſt died, nor could they be revived by expoſing them to the ſun; and, that thoſe "qui n'avoient ſouffert le froid de la "glaciere que pendant peu de tems" were very lively when permitted to make their eſcape.

* In the ſame manner the generical name in other languages, for this tribe of birds, always means the martin, and not the ſwallow.

Thus Anacreon complains of the χελιδὼν for waking him by its twittering.

Now if it be conſidered that there was only the kitchen chimney in a Grecian houſe, it muſt have been the martin which built under the eaves of the window, that was troubleſome to Anacreon, and not the ſwallow.

Ovid alſo ſpeaking of the neſt of the *hirundo*, ſays,

——luteum ſub trabe figit opus.

by which he neceſſarily alludes to the martin, and not the ſwallow.

† Plan de l'ouvrage, p. 15.

Monſ.

Monſ. de Buffon does not, in this account of his experiment, ſtate the time during which the birds were confined; but as the trial muſt have been made in France, the ſwallows which he procured could not be expected to be torpid either in an ice-houſe * or any other place, becauſe the ſeaſon for their being in that ſtate was not yet arrived.

I cannot alſo agree with M. de Buffon that thoſe birds which were ſhut up the longeſt time died through cold, as he ſuppoſes, but for want of food, as he neither ſupplied them with any flies, nor, if he had, could the ſwallows have caught them in the dark: a very ſhort faſt kills theſe tender animals, which are feeding every inſtant when on the wing.

It therefore ſeems not to follow from this, or any other experiment, that ſwallows muſt neceſſarily migrate (as Monſ. de Buffon ſuppoſes) to the coaſt of Senegal.

* The very name of an ice-houſe almoſt ſtrikes one with a chill; I placed, however, a thermometer in one near Hyde Park Corner, on the 23d of November, where it continued 48 hours, and the mercury then ſtood at 43½ by Fahrenheit's ſcale.

This is therefore a degree of cold which ſwallows ſometimes experience whilſt they continue in ſome parts of Europe, without any apparent inconvenience; and it ſhould ſeem that the cold vapours which may ariſe from the included ice, ſink the thermometer only 7 or 8 degrees, as the temperature in approved cellars is commonly from 50 or 51 throughout the year.

Sir William Hamilton informs me, that he hath frequently ſeen ſwallows in the winter between Naples and Puzzuoli, when the weather was warm; as does Mr. Hunter, F. R. S. that he hath obſerved them during the ſame ſeaſon, on the confines of Spain and Portugal. It ſhould ſeem from this, that very mild and warm weather for any continuance always wakes theſe birds from their ſtate of torpidity.

Swallows are ſeen during the ſummer, in every part of Europe from Lapland to the Southern coaſt of Spain; nor is Europe vaſtly inferior in point of ſize to Africa.

If ſwallows therefore retreat to Africa in the winter, ſhould not they be diſperſed over the whole Continent of Africa, juſt as they are over every part of Europe?

But this moſt certainly is not ſo: Dr. Shaw, who was a very good naturaliſt and attended much to the birds in the neighbourhood of Algiers (as appears by his account of that country), makes no mention of any ſuch circumſtance, nor have we heard of it from any other traveller *.

It muſt be admitted indeed, that Herodotus ſpeaking of a part of upper Egypt (which he had never ſeen) ſays, that kites and ſwallows never leave it †; this, however, totally differs from Monſ. Adanſon's account, who informs us that they diſappear in Senegal on the approach of ſummer.

It ſeems to follow therefore, from this ſilence in others, that ſwallows cannot be accommodated for their winter reſidence in any part of that vaſt continent, but in the neighbourhood of Senegal.

But this is not the whole objection to ſuch an hypotheſis.

* It may alſo be obſerved here, that credit is in ſome meaſure given to M. Adanſon's eyeſight, againſt that of all the Engliſh, French, Dutch, Portugueze, and Danes, who have been ſettled not far from Senegal for above a century, many of which have ſpent the greateſt part of their lives there, and whoſe notice, ſwallows ſeen during the winter, muſt have probably attracted.

† Ικτινοι δε και χελιδονες δι ελεος εουσες ουκ απολειπουσι. Euterpe, p. 98. ed. Gale.

If

If the ſwallows of Europe, when they diſappear in thoſe parts, retreat to the coaſt of Senegal, what neceſſarily follows with regard to a Lapland ſwallow?

I will ſuppoſe ſuch a bird to have arrived ſafely at his winter quarters upon the approach of that ſeaſon in Lapland; but he muſt then, according both to Monſ. Adanſon's and de Buffon's account, return to Lapland in the ſpring, or at leaſt ſome other ſwallow from Senegal fill his place *.

Such a bird immediately upon its arrival on the Southern coaſt of Spain would find the climate and food which it deſired to attain, and all proper conveniences for its neſt: what then is to be its inducement for quitting all theſe accommodations which it meets with in ſuch profuſion, and puſhing on immediately over ſo many degrees of European continent to Lapland, where both martin and ſwallow can procure ſo few eaves of houſes to build upon? What alſo is to be the inducement to theſe birds, when they have arrived at that part of the Norwegian coaſt which is oppoſite to the Ferroe iſlands, to croſs degrees of ſea, in order

* Mr. Stephens, A. S. S. informs me, that there was a neſt of martins for twenty years together in the hall of his houſe in Somerſetſhire (near Bath); nor could the old birds procure food either for themſelves, or their young, till the door was opened in the morning.

Can it it be ſuppoſed that the ſame birds or their deſcendants could have ſo long fixed upon ſo very inconvenient a ſpot, to which they conſtantly returned from the coaſt of Africa, neglecting ſo many others, which they muſt have always paſſed by? Does it not alſo afford a moſt ſtrong preſumption, that they were torpid during winter in the neighbourhood of this old hall?

to

to build in ſuch ſmall ſpots of land, where there are ſtill fewer houſes?

The next fact I have happened to meet with of a bird's being ſeen at a conſiderable diſtance from the ſhore, is in Mr. Forſter's lately publiſhed tranſlation of Kalm's account of N. America*.

We are there informed that a bird (which Kalm calls a ſwallow) was ſeen near the ſhip on the 2d of September, and, as he ſuppoſes, 20 degrees from the continent of America †.

It appears however, by what he before ſtates in his journal, that the ſhip was not above 5 degrees from the iſland of Sable.

Beſides, if it is contended that this was an European ſwallow on its paſſage acroſs the Atlantic on the 2d of September, it is too early even for a ſwift, to have been on its migration, which diſappears with us ſooner than the three other ſpecies of European ſwallows ‡.

Only two more inſtances have occurred of birds being ſeen in *open* ſea that have been deſcribed

* Vol. I. p. 24.

† It may not be improper here to obſerve, that in all inſtances of birds being ſeen at ſea any great diſtance from the coaſt, it is not improbable that they may have before ſettled on ſome other veſſel, or perhaps on a piece of floating wreck.

By accidents of this ſort, even butterflies have ſometimes been caught by the ſailors at 40 leagues diſtance from any land. See Monſ. l'Abbé Courte de la Blanchadiere's Voyage to Brazil, Paris, 1759, 21mo. p. 169.

‡ The bird mentioned by Kalm was probably an American ſwallow, forced out to ſea by ſome accidental ſtorm: there are ſeveral ſpecies of them and they ſeem to bear a general affinity to thoſe of Europe.

with

with any ſort of preciſion, which I ſhall juſt ſtate, as I would not decline giving the beſt anſwer I am able to every argument and fact which may be relied upon, by thoſe who contend that birds periodically migrate acroſs oceans.

On the 30th of March, 1751, Oſbeck, in his voyage from Sweden to China *, met with a ſingle houſe ſwallow near the Canary Iſlands, which was ſo tired that it was caught by the ſailors: Oſbeck alſo ſtates, that though it had been fine weather for ſeveral preceding days, the bird was as wet as if it had juſt emerged from the bottom of the ſea.

If this inſtance proves any thing, it is the ſubmerſion and not the migration of ſwallows ſo generally believed in all the northern parts of Europe. It would ſwell this Letter to a moſt unreaſonable ſize, to touch only upon this litigated point; and I ſhall, for the preſent, ſuppreſs what hath happened to occur to me on this controverted queſtion †.

* See the lately publiſhed tranſlation of this voyage.

† I will, however, mention one moſt deciſive fact on this head.

Mr. Stephens, A. S. S. informs me, that, when he was fourteen years of age, a pond of his father's (who was vicar of Shrivenham in Berkſhire) was cleaned, during the month of February; that he picked up himſelf a cluſter of three or four ſwallows (or martins), which were caked together in the mud, and that he carried them into the kitchen, on which they ſoon afterwards flew about the room, in the preſence of his father, mother, and others. Mr. Stephens alſo told me, that his father (who was a naturaliſt) obſerved at the time, he had read of ſimilar inſtances in the northern writers. This fact is alſo confirmed to me by the Reverend Dr. Pye, who was then at ſchool in Shrivenham, as alſo by a very ſenſible land-ſurveyor, who now lives in the village.

Oſbeck afterwards, in the courſe of his voyage, mentions, that a ſwallow (indefinitely) followed the ſhip, near Java, on the 24th of July, and another on the 14th of Auguſt, in the Chineſe ſea, as he terms it.

After what I have obſerved before with regard to other inſtances of the ſame ſort, I need ſcarcely ſay that this naturaliſt does not ſtate of what ſpecies theſe ſwallows were; and that, from the latitudes in which they were ſeen, they muſt have been ſome of the Aſiatic kinds.

I cannot, however, diſmiſs this article of the ſwallow, without adding ſome general reaſons, which ſeem to prove the great improbability of this or any other bird's periodically migrating over wide tracts of ſea; and I the rather do it in this place, becauſe

There are ſeveral reaſons why ſwallows ſhould not be frequently thus found; ponds are ſeldom cleaned in the winter, as it is ſuch cold work for the labourers; and the ſame inſtinct which prompts the bird thus to conceal itſelf, inſtructs it to chooſe ſuch a place of ſecurity, that common accidents will not diſcover it.

But the ſtrongeſt reaſon for ſuch accounts not being more numerous, is, that facts of this ſort are ſo little attended to; for though I was born within half a mile of this pond, and have always had much curioſity with regard to ſuch facts, yet I never heard a ſyllable about this very material and intereſting account, till very lately.

To this fact I muſt alſo add, that ſwallows may be conſtantly taken in the month of October, during the dark nights, whilſt they ſit on the willows in the Thames, and that one may almoſt inſtantaneouſly fill a large ſack with them, becauſe at this time they will not ſtir from the twigs, when you lay your hands upon them. This looks very much like their beginning to be torpid before they hide themſelves under the water.

A man near Brentford ſays, that he hath caught them in this ſtate in the eyt oppoſite to that town, even ſo late as November.

the

the ſwallow is commonly pitched upon as the moſt notorious inſtance of ſuch a regular paſſage.

This ſeems to ariſe firſt from its being ſeen in ſuch numbers during the ſummer, from its appearing almoſt always on the wing, and from its feeding in that poſition; from which two latter circumſtances it is ſuppoſed to be the beſt adapted for ſuch diſtant migrations.

And firſt, let us conſider, from the few facts or reaſons we have to argue from, what length of flight either a ſwallow or any other bird is probably equal to.

A ſwallow, it is true, ſeems to be always on the wing; but I have frequently attended, as much as I could, on a particular one; and it hath appeared to me, that the bird commonly returned to its neſt in eight or ten minutes: as for extent of flight, I believe I may venture to ſay, that theſe birds are ſeldom a quarter of mile from their mate or young ones; they feed whilſt on the wing, and are perpetually turning ſhort round to catch the inſects, who endeavour to elude them as a hare does a greyhound.

It therefore ſeems to me, that ſwallows are by no means equal to long flights, from their practice during their ſummer reſidence with us.

I have long attended to the flight of birds; and it hath always appeared to me, that they are never on the wing for amuſement (as we walk or ride), but merely in ſearch of food.

The only bird which I have ever obſerved to fly without any particular point of direction, is the rook: theſe birds will, when the wind is high,

"Ride in the whirlwind, and enjoy the ſtorm."

They never fly, however, at this time, from point to point, but only tumble in the air, merely for their diverſion.

It ſeems, therefore, that birds are by no means calculated for flights acroſs oceans, for which they have no previous practice: and they are, in fact, always ſo fatigued, that, when they meet a ſhip at ſea, they forget all apprehenſions, and deliver themſelves up to the ſailors.

Let us now conſider another objection to the migration of the ſwallow, which Monſ. de Buffon ſuppoſes may croſs the Atlantic to the Line in eight days *; and this not only from the want of reſt, but of food, during the paſſage.

A ſwallow, indeed, feeds on the wing: but where is it to find any inſects, whilſt it is flying over a wide expanſe of ſea? This bird, therefore, if it ever attempted ſo adventurous a paſſage, would ſoon feel a want of food, and return again to land, where it had met with a conſtant ſupply from minute to minute.

I am aware it may be here objected, that the ſwallow leaves us on the approach of winter, when ſoon no flying inſects can be procured: but I ſhall hereafter endeavour to ſhew, that theſe birds are then torpid, and, conſequently, can want no ſuch food.

Another objection remains to the hypotheſis of migration, which is, that birds, when flying from

* Diſcours ſur la nature des oiſeaux, p. 32.

point

point to point, endeavour always to have the wind againſt them *, as is periodically experienced by the London bird-catchers, in March and October, when they lay their nets for ſinging birds †.

The reaſon, probably, for birds thus flying againſt the wind is, that their plumage may not be ruffled, which indeed I have before had occaſion to mention.

Let us ſuppoſe, then, a ſwallow to be equal to a paſſage acroſs the Atlantic in other reſpects; how is the bird to be inſured of the wind's continuing for days in the ſame quarter; or how is he to depend upon its continuing to blow againſt his flight with moderation? for who can ſuppoſe that a ſwallow can make his way to the point of direction, when buffeted by a ſtorm blowing in the teeth of his intended paſſage ‡?

Laſtly, can it be conceived that theſe, or any other birds, can be impelled by a providential inſtinct, regularly to attempt what ſeems to be attended with ſuch inſuperable difficulties, and what moſt frequently leads to certain deſtruction?

But it will ſtill be objected, that as ſwallows regularly appear and diſappear at certain ſeaſons, it is incumbent upon thoſe who deny their migration, to ſhew

* Kalm, in his voyage to America, makes the ſame obſervation, with regard to flying fiſh, and Valentine ſays, that if the wind does not continue to blow againſt the bird of paradiſe, it immediately drops to the ground.

† Theſe birds, as it ſhould ſeem, are then in motion; becauſe, at thoſe ſeaſons, the ground is plowed either for the winter or lent corn.

‡ I have myſelf attended to ſwallows during a high wind, and have obſerved that they fly only in ſheltered places, whilſt they almoſt touch the ſurface of the ground.

ſhew what becomes of them in Europe during our winter.

Though it might be anſwered, that it is not neceſſary, thoſe who endeavour to ſhew the impoſſibility of another ſyſtem or hypotheſis, ſhould from thence be obliged to ſet up one of their own; yet I ſhall, without any difficulty, ſay, that I at leaſt am convinced ſwallows (and perhaps ſome other birds) are torpid during the winter.

I have not, I muſt own, myſelf ever ſeen them in this ſtate; but, having heard inſtances of their being thus found, from others of undoubted veracity, I have not ſcarcely the leaſt doubt with regard to this point.

It is, indeed, rather difficult to conceive why ſome ornithologiſts continue to withhold their aſſents to ſuch a cloud of witneſſes, except that it perhaps contradicts a favourite hypotheſis which they have already maintained.

Why is it more extraordinary that ſwallows ſhould be torpid during the winter, than that bats are found in this ſtate, and ſo many inſects, which are the food of ſwallows?

But it may be ſaid, that as the ſwallows have crowded the air during the ſummer, in every part of Europe ſince the creation, and as regularly diſappear in winter, why have not the inſtances of their being found in a torpid ſtate been more frequent?

To this it may be anſwered, that though our globe may have been formed ſo many centuries, yet the inhabitants of it have ſcarcely paid any attention to the ſtudy of natural hiſtory, but within theſe late years.

As

As for the ancient Greeks and Romans, their dreſs prevented their being ſo much in the fields as we are; or, if they heard of a rather extraordinary bird in their neighbourhood, they had not a gun to ſhoot it: the only method of attaining real knowledge in natural hiſtory, depends almoſt entirely upon the having frequent opportunities of thus killing animals, and examining them when dead.

If they did not ſtir much in their own country, much leſs did they think of travelling into diſtant regions; want of bills of exchange, and of that curioſity which ariſes from our being thoroughly acquainted with what is near us at home, probably occaſioned this; to which may alſo be added, the want of a variety of languages: ſcarcely any Greek ſeems to have known more than his own tongue, nor Roman more than two *.

Ariſtotle, indeed, began ſomething like a ſyſtem of natural hiſtory, and Pliny put down, in his common place-book, many an idle ſtory; but, before the invention of printing, copies of their works could not be ſo generally diſperſed, as to occaſion much attention to what might be intereſting facts for the natural hiſtorian.

In the ſixteenth century, Geſner, Belon, and Aldrovandus, publiſhed ſome materials, which might be of uſe to future naturaliſts; but, in the ſeventeenth, Ray and Willoughy firſt treated this extenſive branch of ſtudy, with that clearneſs of method,

* It need be ſcarcely here mentioned alſo, that their navigation was confined to the Mediteranean, from the compaſs not having been then diſcovered.

perſpicuity

perſpicuity of deſcription, and accuracy of obſervation, as hath not, perhaps, been ſince exceeded.

The works of theſe great naturaliſts were ſoon diſperſed over Europe, and the merit of them acknowledged; but it ſo happened, that Sir Iſaac Newton's amazing diſcoveries in natural philoſophy making their appearance about the ſame time, engaged entirely the attention of the learned.

In proceſs of time, all controverſy was ſilenced by the demonſtration of the Newtonian ſyſtem; and then the philoſophical part of Europe naturally turned their thoughts to other branches of ſcience.

Since this period, therefore, and not before, natural hiſtory hath been ſtudied in moſt countries of Europe; and conſequently, the finding ſwallows in a ſtate of torpidity, or on the coaſt of Senegal, during the winter, begins to be an intereſting fact, which is communicated to the world by the perſon who obſerves it.

To this I may add, that the common labourers, who have the beſt chance of finding torpid birds, have ſcarcely any of them a doubt with regard to this point; and conſequently, when they happen to ſee them in this ſtate, make no mention of it to others; becauſe they conſider the diſcovery as neither uncommon or intereſting to any one.

Molyneux, therefore, in the Philoſophical Tanſactions *, informs us, that this is the general belief of the common people of Ireland, with regard to land-rails; and I have myſelf received the ſame anſwer from a perſon who, in December, found ſwallows torpid in the ſtump of an old tree.

* Phil. Tranſ. abr. Vol. II. p. 853.

Another

Another reaſon why the inſtances of torpid ſwallows may not be expected ſo frequently, is, that the inſtinct of ſecreting themſelves at the proper ſeaſon of the year, likewiſe ſuggeſts to them, it's being neceſſary to hide themſelves in ſuch holes and caverns, as may not only elude the ſearch of man, but of every other animal which might prey upon them; it is not therefore by any common accident that they are ever diſcovered in a ſtate of torpidity.

Since the ſtudy of natural hiſtory, however, hath become more general, proofs of this fact are frequently communicated, as may appear in the Britiſh Zoology *.

That it may not be ſaid, however, I do not refer to any inſtance which deſerves credit, if properly ſifted, I beg leave to cite the letter from Mr. Achard to Mr. Collinſon, printed in the Philoſophical Tranſactions †, from whence it ſeems to be a moſt irrefragable fact, that ſwallows ‡ are annually diſcovered in a torpid ſtate on the banks of the Rhine. I ſhall alſo refer to Dr. Birch's Hiſtory of the Royal Society ‖, where it is ſtated, that the celebrated Harvey diſſected

* See Vol. II. p. 250. Brit. Zool. ill. p. 13, 14. As alſo Mr. Pennant's Tour in Scotland, p. 199.

† 1763, p. 101.

‡ " Swallows or martins," are Mr. Achard's words, which I the rather mention, becauſe Mr. Collinſon complains that the ſpecies is not ſpecified.

Mr. Collinſon himſelf had endeavoured to prove, that ſand martins are not torpid, Phil. Tranſ. 1760, p. 109. and concludes his letter, by ſuppoſing that all the ſwallow tribe migrates, therefore the ſwift is the only ſpecies remaining; for his friend Mr. Achard ſhews to demonſtration, that ſwallows or martins are torpid; he does not, indeed, preciſely ſtate which of them.

‖ Vol. IV. p. 537.

ſome, which were found in the winter, under water, and in which he could not obſerve any circulation of the blood *.

Aſſuming it, therefore, from theſe facts, that ſwallows have been found in ſuch a ſtate, I would aſk the partiſans of migration, whether any inſtance can be produced where the ſame animal is calculated for a ſtate of torpidity and, at the ſame time of the year, for a flight acroſs oceans?

But it may be urged, poſſibly, that if ſwallows are torpid when they diſappear, the ſame thing ſhould happen with regard to other birds, which are not ſeen in particular parts of the year.

To this I anſwer, that this is by no means a neceſſary inference: if, for example, it ſhould be inſiſted that other birds beſides the cuckow are equally careleſs with regard to their eggs, it would be immediately allowed that the argument ariſing from

* As the ſwallows were found in the winter, they muſt have been in a ſtate of torpidity, as otherwiſe the animals muſt have been putrid.

I ſhall likewiſe here refer to Phil. Tranſ. abr. Vol. V. p. 33. where Mr. Derham ſays, that he heard a ſwift ſqueak in an hole of his houſe on the 17th of April; but that, the weather being cold, it did not ſtir abroad for ſeveral days.

This ſeems to be a ſtrong inſtance of a bird's firſt waking from a ſtate of torpidity, but reſuming its ſleep on the weather being ſevere.

I ſhall cloſe the proofs on this head (which I could much enlarge) by the dignified teſtimony of Sigiſmond, King of Poland, who affirmed on his oath, to the cardinal Commendon, that he had frequently ſeen ſwallows, which were found at the bottom of lakes. See the life of cardinal Commendon, p. 211. Paris, 1671. 4to.

ſuch

ſuch ſuppoſed analogy could by no means be relied upon *.

It is poſſible, however, that ſome other birds, which are conceived to migrate, may be really torpid as well as ſwallows; and if it be aſked why they are not ſometimes alſo ſeen in ſuch a ſtate during the winter, the anſwer ſeems to be, that perhaps there may be a thouſand ſwallows to any other ſort of bird, and that they commonly are found torpid in cluſters.

* I here ſuppoſe the common notion about the cuckow to be true; becauſe both learned and ignorant ſeem equally to agree in the fact.

During the preſent ſummer, however, a girl brought a full feathered young cuckow to a gentleman's houſe, where I happened to be, who ſaid, that it had been for ſeveral days before fed by another bird of equal ſize with itſelf; which therefore could not be a hedge-ſparrow, or other ſmall bird, but the parent cuckow.

I have alſo lately been favoured, by Mr. Pennant, with the following extract from a manuſcript of Derham's on inſtinct.

" The Rev. Mr. Stafford was walking in Gloſſop-dale in the " Peak of Derbyſhire, and ſaw a cuckow riſe from its neſt, " which was on the ſtump of a tree, that had been ſome time " felled, ſo as much to reſemble the colour of the bird. In " this neſt were two young cuckows, one of which he " faſtened to the ground, by means of a peg and line, and very " frequently, for many days, beheld the old cuckow feed theſe " her young ones."

It is not impoſſible, therefore, that this moſt general opinion will turn out like the ſuppoſed effects of the venom of the tarantula; and, indeed, it is difficult to conceive how ſo ſmall a bird as a hedge-ſparrow can feed a cuckow: it is alſo remarkable, that the witneſſes often vary about the ſpecies of ſmall bird thus employed.

It is poſſible, however, that the cuckow (though it may not hatch its young) may feed them, when grown too large for the foſter parent.

If a ſingle bird of any other kind happens to be ſeen in the winter, without motion or apparent warmth, it is immediately conceived that it died by ſome common accident.

I ſhall, however, without any reſerve, ſay, that I rather conceive the notion which prevails with regard to the migration of many birds, may moſt commonly ariſe from the want of obſervation, and ready knowledge of them, when they are ſeen on the wing, even by profeſſed ornithologiſts.

It is an old ſaying, that "a bird in the hand is "worth two in the buſh;" and this holds equally with regard to their being diſtinguiſhed, when thoſe even who ſtudy natural hiſtory, have but a tranſient ſight of the animal *.

If, therefore, a bird, which is ſuppoſed to migrate in the winter, paſſes almoſt under the noſe of a Linnæan, he pays but little attention to it, becauſe he cannot examine the beak, by which he is to claſs the bird. Thus I conceive, that the ſuppoſing a nightingale to be a bird of paſſage ariſes from not readily diſtinguiſhing it, when ſeen in a hedge, or on the wing †.

This bird is known to the ear of every one, by its moſt ſtriking and capital notes, but to the eye of very

* An ingenious friend of mine makes always a very proper diſtinction between what he calls in-door and out-door naturaliſts.

Thomas Williſel, who aſſiſted Ray and Willughby much with regard to the natural hiſtory of the animals of this iſland, never ſtirred any where without his gun and fiſhing-tackle.

† No two birds fly in the ſame manner, if their motions are accurately attended to.

few indeed; becaufe the plumage is dull, nor is there any thing peculiar in its make.

The nightingale fings perhaps for two months *, and then is never heard again till the return of the fpring, when it is fuppofed to migrate to us from the continent, with redftarts, and feveral other birds.

That it cannot really do fo, feems highly probable, from the following reafons.

This bird is fcarcely ever feen to fly above twenty yards, but creeps at the bottom of the hedges, in fearch of maggots, and other infects, which are found in the ground.

If the fwallow is not fupplied with any food during its paffage acrofs oceans, much lefs can the nightingale be fo accommodated; and I have great reafon to believe, from the death of birds in a cage, which have had nothing to eat for twenty-four hours, that thefe delicate and tender animals cannot fupport a longer faft, though ufing no exercife at all.

To this I may alfo add, that thofe birds which feed on infects are vaftly more feeble than thofe whofe bills can crack feed, and confequently, lefs capable of bearing any extraordinary hardfhips or fatigue.

But other proofs are not wanting, that this bird cannot migrate from England.

* Whilft it fings even, the bird can feldom be diftinguifhed, becaufe it is then almoft perpetually in hedges, when the foliage is thickeft, upon the firft burft of the fpring, and when no infects can as yet have deftroyed confiderable parts of the leaves.

Nightin-

Nightingales are very common in Denmark, Sweden, and Ruſſia *, as alſo in every other part of Europe, as well as Aſia, if the Arabic name is properly tranſlated.

Now, if it is ſuppoſed that many of theſe birds which are obſerved in the ſouthern parts of England, croſs the German ſea, from the oppoſite coaſt of the continent; why does not the ſame inſtinct drive thoſe of Denmark to Scotland, where no ſuch bird was ever ſeen or heard † ?

But theſe are not all the difficulties which attend the hypotheſis of migration; nightingales are agreed to be ſcarcely ever obſerved to the weſtward of Dorſetſhire, or in the principality of Wales ‡, much leſs in Ireland.

I have alſo been informed, that theſe birds are not uncommon in Worceſterſhire, whereas they are exceſſively rare (if found at all) in the neighbouring county of Hereford.

Whence, therefore, can it ariſe, that this bird ſhould at one time be equal to the croſſing of ſeas, and at other times not travel a mile or two into an adjacent county? Does it not afford, on the other hand, a ſtrong proof, that the bird really continues

* See Dr. Birch's Hiſtory of the Royal Society, Vol. III. p. 189. Linnæi Fauna Suecica. and Biographia Britannica, art. FLETCHER; where it is ſaid, that they have in Ruſſia a greater variety of notes than elſewhere.

† Sir Robert Sibbald, indeed, conceives the nightingale to be a bird of North Britain; but, if I can depend upon many concurrent teſtimonies, no ſuch bird is ever ſeen or heard ſo far northward at preſent, nor could I ever trace them in that direction further than Durham.

‡ I have, however, frequently ſeen the nightingale's congener (and ſuppoſed fellow-traveller) the redſtart in Wales.

on

on the ſame ſpot during the whole year, but happens not to be attended to, from the reaſons I have before ſuggeſted?

I am therefore convinced, that if I was ever to live in the country during the winter, I ſhould ſee nightingales, becauſe I ſhould be looking after them, and I am accordingly informed, by a perſon who is well acquainted with this bird, that he hath frequently obſerved them during this ſeaſon *.

If it be aſked, why the nightingales are all this time mute? the anſwer is, that the ſame ſilence is experienced in many other birds, and this very muteneſs is, in part the cauſe why the bird is not attended to in winter.

I muſt now aſk thoſe who contend for the migration of a nightingale, what is to be its inducement for croſſing from the continent to us? a ſwallow, indeed, may want flies in winter, if it ſtays in England; but a nightingale is juſt as well ſupplied with inſects on the continent, as it can be with us after its paſſage †. I muſt alſo aſk, in what other part of the

* I find they have alſo been ſeen in France during the winter. See a treatiſe, intitled, Aëlologue, Paris 1751. p. 23.

† I have omitted the mention of a more minute proof, that this bird cannot migrate from the continent, from the having kept them for ſome years in a cage, and having been very attentive to their ſong.

Kircher (in his Muſurgia) hath given us the nightingale's notes in muſical characters, from which it appears that the ſong of a German nightingale differs very materially from that of an Engliſh one: now, if there was a communication by migration between the continent and England, the ſong of theſe birds would not ſo materially differ, as I may, perhaps, ſhew, by ſome experiments I have made, in relation to the notes of birds.

I have before mentioned, that Mr. Fletcher, who was embaſſador from England to Ruſſia in the time of Queen Elizabeth, the world this bird is feen during the winter? muft it migrate to Senegal with the fwallow?

I am perfuaded likewife, that the cuckow never migrates from this ifland any more than the nightingale: this bird is either probably torpid in the winter, or otherwife is miftaken for one of the fmaller kind of hawks *; which it would be likewife in the fpring, was it not for its very particular note at that time, and which only lafts during courtfhip, as it does with the quail.

If there is fine weather in February, this bird fometimes makes this fort of call to its mate, whilft it is fuppofed to continue ftill on the continent.

An inftance is mentioned by Mr. Bradley †, of not only a fingle cuckow, but feveral, which were heard in Lincolnfhire, during the month of February; and that able naturalift Mr. Pennant informs me, another was heard near Hatcham in Shropfhire, on the 4th of February in the prefent year ‡.

obferved that the fong of the Ruffian nightingale differed from that of the Englifh.

* Mr. Hunter, F. R. S. informs me, that he hath feen cuckows in the ifland of Belleifle during the winter, which is not fituated fo much to the fouthward, as to make it improbable that they may equally continue with us.

† Works of Nature, p. 77.

‡ Mr. Pennant received this account from Mr. Plimly, of Longnor in Shropfhire.

Thus likewife Mr. Edwards informs us, that the fea fowls near the Needles, which are commonly fuppofed to migrate in winter, appear upon the weather's being very mild. Effays, p. 197.

It

It is amazing how much the being intereſted to diſcover particular objects contributes to our readily diſtinguiſhing them.

I remember the being much ſurprized that a grey-headed game-keeper always ſaw the partridge on the ground before they roſe, when I could not do the ſame. He told me, however, that the reaſon was, I lived in a time when the ſhooter had no occaſion to give himſelf that trouble.

He then further explained himſelf, by ſaying, that when he was young, no one ever thought of aiming at a bird when on the wing, and conſequently they were obliged to ſee the game before it was ſprung. He added, that from this neceſſity he could not only diſtinguiſh partridges, but ſnipes and woodcocks, on the ground.

Another inſtance of the ſame kind, is the great readineſs with which a perſon, who is fond of courſing, finds a hare ſitting in her form: thoſe, however, who are not intereſted about ſuch ſport, can ſcarcely ſee the hare, when it is under their noſe, and pointed out to them.

But more apparent objects eſcape our notice, when we are not intereſted about them.

Aſk any one, who hath not a botanical turn, what he hath ſeen in paſſing through a rich meadow, at the time it is moſt enamelled with plants in flower; and he will tell you, that he hath obſerved nothing but graſs and daiſies. If moſt gardeners even are in like manner aſked whether the flowers of a bean grow on every ſide of the ſtalk, they will ſuppoſe that they do,

 whereas

whereas they, in reality, are only to be found on one side.

The mouths of flounders are often turned different ways, which one would think could not well escape the observation of the London fishmongers; yet, upon asking several of them whether they had attended to this particular, I found they had not, till I shewed them the proof in their own shops.

A fishmonger, however, knows immediately whether a fish is in good eating order or not, on the first inspection; because this is a circumstance which interests him.

I shall, however, by no means suppress two arguments in favour of migration, which seem to require the fullest answer that can be given to them.

The first is, that there are certain birds, which appear during the winter, but disappear during the summer; and it may be asked, where such birds can be supposed to breed, if they do not migrate from this island.

These birds are in number four, viz. the snipe, woodcock, redwing, and fieldfare.

As for the snipe, I have a very short answer to give to the objection, as far as it relates to this bird; because it constantly breeds in the fens of Lincolnshire, Wolmar forest, and Bodmyn downs; it is therefore highly probable, that it does the same in almost every county of England.

I must own, however, that, till within these few years, I conceived the nest of a snipe was as rarely seen in England, as that of a woodcock or fieldfare; and that able ornithologist Mr. Edwards supposes this to be be the fact, in the late publication of his ingenious Essays on Natural History *.

Woodcocks likewise are known to build in some parts of England every year; but, as the instances are commonly those of a single nest, I would by no means pretend to draw the same proof against the summer migration of this bird, as in the former case of the snipe.

I will most readily admit, that these accidental facts are rather to be accounted for, perhaps, from the whimsy or silliness of a few birds, which occasions their laying their eggs in a place where they are easily discovered, and contrary to what is usual with the bulk of the species.

I remember to have seen a duck's nest once on the top of a pollard willow, near the decoy in St. James's Park; it would not be, however, fair to infer from such an instance, that all ducks would pitch upon the same very improper situation for a nest, upon which it is difficult to conceive how a web-footed bird could settle.

Some silly birds likewise now and then choose a place for building, which cannot escape the observation of either man or beast, as he passes by.

I therefore suppose that the few proofs of woodcocks nests having been found in England, arise either from one or other of these two causes, and all which they seem to prove is, that our climate in summer is not absolutely improper for them.

It is to be observed, however, that Mr. Catesby considers such instances as of equal force against the

* P. 72.

migration of the woodcock, as of the ſnipe *. Willughby alſo ſays, that Mr. Jeſſop ſaw young woodcocks ſold at Sheffield (which rather implies a certain number being brought to market), and that others had obſerved the ſame elſewhere †.

We are, indeed, informed by Scopoli ‡, that they breed conſtantly in Carniola, which is conſiderably to the ſouthward of any part of England: our country is therefore certainly not too hot for them.

Woodcocks appear and diſappear almoſt exactly about the ſame time in every part of Europe, and perhaps Africa ‖: heat and cold, therefore, ſeem not to have any operation whatſoever with regard to the ſuppoſed migration of this bird.

But it may be ſaid, what ſignifies proving the probability of woodcocks breeding in England, if it is not a known fact that they do ſo?

To this it ſhould ſeem there are ſeveral anſwers, as it is equally incumbent upon thoſe who contend for migration, to ſhew that theſe birds were ever ſeen on ſuch paſſage.

Another anſwer is, aſk ninety-nine people out of a hundred, whether ſnipes ever make a neſt in England; and they will immediately ſay, that they do not; ſo little are facts or obſervations of this ſort attended to.

But I ſhall now endeavour to give ſome other reaſons why woodcocks may not only continue with us,

* Phil. Tranſ. abr. Vol. II. p. 889.

† B. iii. c. 1.

‡ Ornith. Leipſig, 1769.

‖ Shaw's Trav. Phyſ. Obſ. ch. ii.

during

during the ſummer, but alſo breed in large tracts of wood or bog, without being obſerved.

In the other parts of Europe, all birds almoſt are conſidered as game, or, at leaſt, are eaten as wholeſome food. Ray therefore mentions, that hawks and owls are ſold by the poulterers at Rome; every ſort of ſmall bird alſo is equally the foreign fowler's object *.

An Engliſhman does not conſider, on the other hand, perhaps twelve kinds of birds worthy his attention, or expence of powder, none of which are ever ſhot in our woods during the ſummer, nor are birds then diſturbed by felling either coppice or timber.

But it will be ſaid, why are not woodcocks ſometimes ſeen, however, as they may be ſuppoſed to leave their cover in ſearch of food?

To this I anſwer, that woodcocks ſleep always in the daytime, whilſt with us in the winter, and feed only during the night †. Whenever a woodcock, therefore, is fluſhed, he is rouſed from his ſleep by the ſpaniel or ſportſman, and then takes wing, becauſe there are no leaves on the trees to conceal the bird.

Whoever hath looked attentively at a woodcock's eye, muſt ſee that, from the appearance of it, the

* In one of Boccace's Novels, a lover, who lives at Florence, dreſſes a falcon for the dinner of his miſtreſs. Giorna a V. Novel. IX.

† Almoſt all the wild fowl of the duck kind alſo ſleep in the daytime, and feed at night.

ſight

fight muſt be more calculated to diſtinguiſh objects by night than by day *.

The fact therefore is notorious to thoſe who cut glades in their woods, and fix nets for catching theſe birds, that they never ſtir but as it begins to be dark, after which they return again by day-break, when their ſight even then is ſo indifferent, that they ſtrike againſt the net, and thus become entangled.

No one with us ever thinks of fixing or attending ſuch nets in ſummer for woodcocks, becauſe it is not then ſuppoſed that there is any ſuch bird in the iſland; if they tried this experiment, however, I muſt own that I believe they would have ſport †.

Mr. Reinhold Forſter, F. R. S. who is an able naturaliſt, informs me, that the fowlers in the neighbourhood of Dantzick kill many woodcocks about St. John's day (or Midſummer), in the following manner,

* I conceive alſo, it is from the eyes looking ſo dull, that this bird is generally conſidered as being ſo fooliſh: hence the Africans call the woodcock *hammar el hadgel*, or the partridge's aſs. Shaw's Phyſ. Obſ. ch. ii.

† I would aſk thoſe who will probably laugh at the very idea of ſuch ſport (which I do not, however, abſolutely inſure), whether, if I was to ſend them to any part of the Britiſh coaſt to catch the true anchovy, or tunny fiſh, they would not ſuppoſe equally that it was a fool's errand.

Notwithſtanding, however, this incredulity, I can produce the authority of both Ray (Syn. Piſc. p. 107.) and Mr. Pennant (Brit. Zool. ill. p. 34. 36.), that the true anchovy is caught in the ſea not far from Cheſter, and the tunny fiſh on the coaſt of Argyleſhire, together with the herrings, where they are called *mackrel ſture*.

Is it not amazing, however, that a fiſh of ſuch a ſize as the tunny ſhould never have been heard of, even by the Scotch naturaliſt Sir Robert Sibbald?

ner, and that they continue to do ſo till the month of Auguſt.

They wait on the ſide of ſome of the extenſive woods in that neighbourhood, before day-break, for the return of the woodcock from his feeding in the night-time, and always depend upon having a very good chance of thus ſhooting many of them.

The Dantzickers, however, might be employed the whole ſummer near theſe woods in the day-time, without ever ſeeing ſuch a bird; and it ſeems therefore not improbable, that it ariſes from our not waiting for them at twilight or day-break, that they are never obſerved by Engliſhmen in the ſummer. If this bird ſhould, however, be ſeen in the night, it is immediately ſuppoſed to be an owl, which a woodcock does not differ much from in its flight.

To theſe reaſons for woodcocks not being obſerved, it may be added, that the bird is believed to be abſolutely mute, and conſequently, never diſcovers itſelf by its call.

If it be ſtill contended, that the neſt or young muſt ſometimes be ſtumbled upon, though in the centre of extenſive woods, or large bogs, the ſiſkin (or aberdavine *) is a much more extraordinary inſtance of concealing its neſt and young.

The plumage of this bird is rather bright than otherwiſe; and the ſong, though not very pleaſing, yet is very audible, both which circumſtances ſhould diſcover it at all times; yet Kramer †, informs us, that, though immenſe numbers breed annually on

* Brit. Zool. p. 309.

† Elenchus Animalium per Auſtriam, p. 261. Viennæ, 1756.

the banks of the Danube, no one ever obſerved the neſt.

This bird is rather uncommon in England; ſo that if I aſk when the neſt was ever found within the verge of the iſland, it may be conſidered as rather an unfair challenge.

There is another bird, however, called a red-poll *, which is taken in numbers during the Michaelmas and March flights by the London bird-catchers, whoſe neſt, I believe, was never diſcovered in England, though I have ſeen them in pairs during the ſummer, both in the mountainous parts of Wales and highlands of Scotland †.

But I ſhall now mention another proof that woodcocks breed in England.

The Reverend Mr. White, of Selborn, who is not only a well-read naturaliſt, but an active ſportſman, informs me, that he hath frequently killed woodcocks in March, which, upon being opened, had the rudiments of eggs in them, and that it is uſual at that time to fluſh them in pairs. Willughby alſo obſerves the ſame ‡.

This bird, therefore, certainly pairs before its ſuppoſed migration; and can it be conceived that this ſtrict union (which birds in a wild ſtate ſo faithfully adhere to) ‖, ſhould take place before they

* Brit. Zool. p. 312.

† This elegant little bird is very common in Hudſon's Bay, where it feeds chiefly on the birch trees; which being more common in the northern than ſouthern parts of Great Britain, may account for the bird's being more often ſeen northward.

‡ B. III. c. i.

‖ It is believed that no mule-bird was ever ſeen in a wild ſtate, notwithſtanding M. de Buffon ſuſpects many an intrigue

traverſe

traverſe oceans, and when they cannot as yet have pitched upon a proper place for concealing their neſt and neſtlings?

Let us examine if this intercourſe before migration takes place in other birds, which are ſuppoſed to croſs wide extents of ſea: and a quail affords ſuch proof.

I have been preſent when theſe birds have been caught in the ſpring, which always turn out to be males, and are enticed to the nets by the call of the hen; quails therefore pair after they appear in England.

But I ſhall now conſider the other two inſtances of birds which are ſeen with us in the winter, and are not obſerved in the ſummer; I mean, the fieldfare and redwing.

And firſt, let us examine, where theſe birds are actually known to breed: the northern naturaliſts ſay, in Sweden; Klein, in the neighbourhood of Dantzick, which is only in lat. 54° 30′ *; and Willughby, in Bohemia.

in the receſſes of the woods" (Hiſt. Nat. des Oiſeaux, tom. I.) ſuch irregular intercourſe is only obſerved in cages and aviaries, where birds are not only confined, but pampered with food.

* See Klein, de Avibus Erraticis, p. 178. Klein, however, cites Zornius, who lived in the ſame part of Germany, and who aſſerts that the *turdus Iliacus* (or redwing) leaves thoſe parts in the ſpring. The circumſtance therefore of the redwing's breeding in numbers *(per multitudines)* had eſcaped the notice of Zornius, though he hath written a diſſertation on this queſtion.

Is it at all ſurprizing, after this, that ſuch diſcoveries, if made at all, ſhould not be commonly heard of?

As they therefore build their nests in more Southern parts of Europe, there is certainly no natural impossibility of their doing so with us, though, I must own, I never yet heard but of one instance, which was a fieldfare's nest found near Paddington *.

I cannot, however, but think it is only from want of observation, that more of such nests have not been discovered, which are only looked after by very young children; and the chief object is the eggs, or nestlings, not the bird which lays them †.

The plumage therefore and flight of the fieldfare or redwing being neither of them very remarkable, it is not at all improbable they may remain in summer, without being attended to; and particularly the redwing, which scarcely differs at all in appearance from other thrushes. Thus the cough is by no means peculiar to Cornwall, as is commonly supposed, but is mistaken for the jackdaw, or rook.

But it may be said, that these birds fly in flocks during the winter, and if they remain here during the summer, we should see them equally congregate.

I have not before referred to Klein, who hath written a very able treatise, in which he argues against the possibility of migration in birds; because, though I should be very happy to support my poor opinion by his authority, yet I thought it right neither to repeat his facts, or arguments.

* See also Harl. Misc. Vol. II. p. 561.

† Many birds also build in places of such difficult access, that boys cannot climb to; birds-nesting is confined almost entirely to hedges, and low shrubs.

This circumſtance, however, is by no means peculiar to the fieldfare and redwing; moſt of the hard-billed ſinging birds do the ſame in winter, but ſeparate in ſummer, as it is indeed neceſſary all birds ſhould during the time of breeding.

I ſhall now conſider another argument in favour of migration, which I do not know hath been ever inſiſted upon by thoſe writers who have contended for it, and which at firſt appearance ſeems to carry great weight with it.

There are certain birds, which are ſuppoſed to viſit this iſland only at diſtant intervals of years; the Bohemian chatterer and croſs-bill * (for example) once perhaps in twenty.

The fact is not diſputed, that ſuch birds are not commonly obſerved in particular ſpots from year to year; but this may ariſe from two cauſes, either a partial migration within the verge of our iſland, or perhaps more frequently from want of a ready knowledge of birds on the wing, when they happen to be ſeen indeed, but cannot be examined.

I never have diſputed ſuch a partial migration; and indeed I have received a moſt irrefragable proof of ſuch a flitting, from the Rev. Mr. White of Selborn in Hampſhire, whoſe accurate obſervations I have before had occaſion to argue from.

* This bird changes the colour of its plumage at different ſeaſons of the year, which is ſometimes red.

The firſt account we have of their being ſeen, is in the Ph. Tr. abr. Vol. V. p. 33. where Mr. Edward Lhwyd ſuſpects them to be Virginia nightingales, from their feathers being red, and had no difficulty of at once ſnppoſing that they had croſſed the Atlantic.

The rock (or ring-ouzel) hath always hitherto been conſidered as frequenting only the more mountainous parts of this iſland: Mr. White, however, informs me that there is a regular migration of theſe birds, which flock in numbers, and regularly viſit the neighbourhood of Selborn, in Hampſhire *.

I therefore have little doubt but that they equally appear in others of our Southern counties; though it eſcapes common obſervation, as they bear a ſort of general reſemblance to the black-bird, at leaſt to the hen of that ſpecies.

I own alſo, that I always conceived the Bohemian chatterer was not obſerved in Great Britain but at very diſtant intervals of years, and then perhaps only a ſingle bird, whereas Dr. Ramſey (profeſſor of natural hiſtory at Edinburgh) informs Mr. Pennant, that flocks of theſe birds appear conſtantly every year in the neighbourhood of that city †.

As for croſs-bills, they are ſeen more and more in different parts of England, ſince there have been ſo many plantations of firs: this bird is remarkably fond of the ſeeds of theſe trees, and therefore changes its place to thoſe parts where it can procure the greateſt plenty of ſuch food ‡.

* See alſo Br. Zool. Ill. p. 56.

† Theſe birds are ſaid to be particularly fond of the berries of the mountain-aſh, which is an uncommon tree in the Southern parts of Great Britain, but by no means ſo in the North.

‡ This bird ſhould alſo, for the ſame reaſon, be found from year to year in the cyder counties, if it was true (as is commonly ſuppoſed) that he is particularly fond of the kernels of

This flitting therefore by no means amounts to a total and periodical migration over ſeas, but is no more than what is experienced with regard to ſeveral birds.

For example, the Britiſh Zoology informs us *, that, at an average, 4000 dozen of larks are ſent up from the neighbourhood of Dunſtable, to ſupply the London markets; nor do I hear, upon inquiry, that there is any complaint of the numbers decreaſing from year to year, notwithſtanding this great conſumption.

I ſhould not ſuppoſe that 50 dozen of ſkylarks are caught in any other county of England; and it ſhould therefore ſeem that the larks from the more adjacent parts croud in to ſupply the vacuum occaſioned by the London Epicures, which may be the cauſe poſſibly of a partial migration throughout the whole iſland.

I begin now to approach to ſomething like a concluſion of this (I fear) tedious diſſertation: I think, however, that I ſhould not omit what appears to me at leaſt as a demonſtration, that one bird, which is commonly ſuppoſed to migrate acroſs ſeas, cannot poſſibly do ſo.

apples, which it is conceived he can inſtantly extract with his very ſingular bill.

Mr. Tunſtall, F. R. S. however, at my deſire, once placed an apple in the cage of a croſs-bill, which he had kept for ſome time in his very valuable and capital collection of live birds: upon examining the apple a fortnight afterwards, it remained untouched.

* P. 235.

A landrail

A landrail *, when put up by the ſhooter, never flies 100 yards; its motion is exceſſively ſlow, whilſt the legs hang down like thoſe of the water fowls which have not web feet, and which are known never to take longer flights.

This bird is not very common with us in England, but is exceſſively ſo in Ireland, where they are called corn-creaks.

Now thoſe who contend that the landrail, becauſe it happens to diſappear in winter, muſt migrate acroſs oceans, are reduced to the following dilemma.

They muſt firſt either ſuppoſe that it reaches Ireland periodically from America; which is impoſſible, not only becauſe the paſſage of the Atlantic includes ſo many degrees of longitude, but becauſe there is no ſuch bird in that part of the globe.

If the landrail therefore migrates from the continent of Europe to Ireland, which it muſt otherwiſe do, the neceſſary conſequence is, that many muſt paſs over England in their way Weſtward to Ireland; and why do not more of theſe birds continue with us, but, on the contrary, immediately proceed acroſs the St. George's channel?

Whence ſhould it ariſe alſo, if they paſs over this iſland periodically in the ſpring and autumn, that they are never obſerved in ſuch paſſage, as I have already ſtated their rate in flying to be exceſſively ſlow; to which I may add, that I never ſaw them riſe to the height of twenty yards from the ground, nor indeed exceed the pitch of a quail.

* Br. Zool. p. 387.

I have

I have now submitted the best anſwers that have occurred, not only to the general arguments for the migration of birds acroſs oceans, but alſo to the particular facts, which are relied upon as actual proofs of ſuch a regular and periodical paſſage.

Though I may be poſſibly miſtaken in many of the conjectures I have made, yet I think I cannot be confuted but by new facts, and to ſuch freſh evidence, properly authenticated, I ſhall moſt readily give up every point, which I have from preſent conviction been contending for.

I may then perhaps alſo flatter myſelf, that the having expreſſed my doubts with regard to the proofs hitherto relied upon, in ſupport of migration, may have contributed to ſuch new, and more accurate obſervations.

It is to be wiſhed, however, that theſe more convincing and deciſive facts may be received from iſlanders (the more diſtant from any land the better*) and not from the inhabitants of a continent; as it does not ſeem to be a fair inference, becauſe certain birds leave certain ſpots at particular times, that they therefore migrate acroſs a wide extent of ſea.

For example, ſtorks diſappear in Holland during the winter, and they have not a very wide tract of ſea between them and England; yet this bird never frequents our coaſts.

* I would particularly propoſe the iſlands of Madera and St. Helena; to theſe, I would alſo add the iſland of Aſcenſion (had it any inhabitants), as likewiſe Juan Fernandez, for the Pacifick ocean.

The

The ſtork, however, may be truely conſidered as a bird of paſſage, by the inhabitants of thoſe parts of Europe (wherever ſituated) to which it may be ſuppoſed to reſort during the winter, and where it is not ſeen during the ſummer.

I am, dear Sir,

Your moſt faithful,

humble ſervant,

Daines Barrington.

P. S.

SINCE I ſent to you my very long letter on the migration of birds, I have had an opportunity of examining the " Planches Enluminées," which are ſaid to be publiſhed under M. de Buffon's inſpection, and which ſeem to afford a demonſtration of M. Adanſon's inaccuracy in ſuppoſing either the roller, or ſwallows, which he caught in his ſhip, near the coaſt of Senegal, to be the ſame with thoſe of Europe.

In the 8th of theſe plates, there is a coloured figure of a bird, called le rollier d'Angola, which agrees exactly with M. Adanſon's deſcription *; but he truſted too much to his memory, when he pronounced it to be the ſame with the Garrulus Argentoratenſis of Willughby, and therefore ſuppoſed it to be on its paſſage to Europe.

This bird hath, indeed, in many reſpects, a very ſtrong reſemblance to the common roller of Europe, which is repreſented alſo in the Planches Enluminées, plate 486; but it differs moſt materially in the length of the two exterior feathers of the tail, as well as in the colour of the neck, which in the African roller is of a moſt bright green, and in the European of rather a dull blue.

In the 310th plate, there is likewiſe a coloured repreſentation of the " Hirondelle a ventre roux du " Senegal," which ſpecimen was poſſibly furniſhed by Monſ. Adanſon himſelf.

* Voyage au Senegal, p. 15. There is alſo another African bird, repreſented in the " Planches Enluminées," which might very eaſily, on a haſty inſpection, be miſtaken for the Garrulus Argentoratenſis, viz. the Guepier a longue queue du Senegal. Pl. Enl. p. 314.

The roller of Angola is alſo engraved by Briſſon, T. ii. pl. 7.

It very much resembles the European ſwallow, but the tail differs, as the forks (in the Senegal ſpecimen) taper from the top of the two exterior feathers to the bottom, at three regular diviſions, whereas in the European they are nearly of the ſame width throughout.

The convincing proof, however, that the "Hirondelle a ventre roux du Senegal" differs from our chimney ſwallow is, that the rump is entirely covered with a bright orange or cheſnut, which in the European ſwallow "is of a very lovely but dark purpliſh blue colour *."

Having lately looked into Ariſtotle's Natural Hiſtory, with regard to the cuckow, I take this opportunity alſo of enlarging on the doubts I have thrown out, in relation to the prevailing notion of this bird's neſtlings being hatched and fed by foſter parents.

I find that this moſt general opinion takes its riſe from what is ſaid by this father of natural hiſtory, in his ninth book, and twenty-ninth chapter.

Ariſtotle there aſſerts, that the cuckow does not build a neſt itſelf, but makes uſe moſt commonly of thoſe of the wood-pigeon, hedge-ſparrow, lark, (which he adds are on the ground) as well as that of the χλωρις †, which is in trees.

Now, if we take the whole of this account together, it is certainly not to be depended upon; for the wood-pigeon ‡ and hedge-ſparrow do not build upon the ground, and no one ever pretended to have

* See Willughby, p. 312.

† The χλωρις is rendered *luteola*; but, as there is no deſcription, it is difficult to ſay what bird Ariſtotle here alludes to; Zinanni ſuppoſes it to be the greenfinch.

‡ The wood-pigeon, from its ſize, ſeems to be the only bird which is capable of hatching, or feeding, the young cuc-

found a cuckow's egg in the neſt of a lark, which, indeed, is ſo placed.

I have before obſerved, that the witneſſes often vary with regard to the bird in which the cuckow's egg is depoſited *; and Ariſtotle himſelf, in the ſeventh chapter of his ſixth book, confines the foſter-parents to the wood-pigeon and hedge-ſparrow, but chiefly the former.

If the age † of Ariſtotle is conſidered, when he began to collect the materials for his Natural Hiſtory, by the encouragement of Alexander after his conqueſts in India ‡, it is highly improbable he ſhould have written from his own obſervations. He therefore ſeems to have haſtily put down the accounts of the perſons who brought him the different ſpecimens from moſt parts of the then known world.

Inaccurate, however, and contradictory as theſe reports often turn out, it was the beſt compilation which the ancients could have recourſe to; and Pliny

kow; yet, if it is recollected that this bird lives on ſeeds, it is probable that the cuckow, whoſe nouriſhment is inſects, would either be ſoon ſtarved, or incapable of digeſting what was brought by the foſter-parent. This objection is equally applicable to the χλωρις, if it is our greenfinch.

* Thus Linnæus ſuppoſes it (in the Fauna Suecica) to be the white wagtail, which bird builds in the banks of rivers, or roofs of houſes, (See Zinanni, p. 51.) where it is believed no young cuckow was ever found.

† He did not leave the ſchool of Plato till the age of thirty-eight (or, as ſome ſay, forty); after which, ſome years paſſed before he became Alexander's preceptor, who was then but fourteen: nor could he have written his Natural Hiſtory, probably, till twelve years after this, as Pliny ſtates that ſpecimens were ſent to him by Alexander, from his conqueſts in India. Ariſtotle therefore muſt have been nearly ſixty, when he began this great work, and conſequently muſt have deſcribed from the obſervations of others.

‡ Pliny, L. viii. c. 16.

therefore professes only to abridge him, in which he often does not do justice to the original.

Whatever was asserted by Aristotle, is well known to have been most implicitly believed, till the last century; and I am convinced that many of the learned in Europe would, before that time, not have credited their own eyesight against what he had delivered.

There cannot be a stronger proof that the general notion about the cuckow arises from what is laid down by Aristotle, than the chapter which immediately follows, as it relates to the goatsucker, and states that this bird sucks the teats of that animal.

From this circumstance, the goatsucker hath obtained a similar name in most languages, though it is believed no one (who thinks at all about matters of this sort) continues to believe that this bird sucks the goat *, any more than the hedgehog does the cow.

I beg leave, however, to explain myself, that I give these additional reasons only for my doubting with regard to this most prevailing opinion; because I am truly sensible that many things happen in nature, which contradict all arguments from analogy, and I am persuaded, therefore, that the first person who gave an account of the flying fish, was not credited by any one, though the existence of this animal is not now to be disputed.

All that I mean to contend for is, that the instances of such extraordinary peculiarities in animals, should be proportionably well attested, in all the necessary circumstances.

I must own, for example, that nothing short of the following particulars will thoroughly satisfy me on this head.

* See Zinanni p. 95. who took great pains to detect this vulgar error.

The

The hedge-ſparrow's neſt muſt be found with the proper eggs in it, which ſhould be deſtroyed by the cuckow, at the time ſhe introduces her ſingle egg *.

The neſt ſhould then be examined at a proper diſtance from day to day, during the hedge-ſparrow's incubation, as alſo the motions of the foſter parent attended to, particularly in feeding the young cuckow, till it is able to ſhift for itſelf.

As I have little doubt that the laſt mentioned circumſtance will appear deciſive to many, without the others which I have required, it may be proper to give my reaſons, why I cannot conſider it alone, as ſufficient.

There is ſomething in the cry of a neſtling for food, which affects all kinds of birds, almoſt as much as that of an infant, for the ſame purpoſe, excites the compaſſion of every human hearer †.

I have taken four young ones from a hen ſkylark, and placed in their room five neſtling nightingales, as well as five wrens, the greater part of which were reared by the foſter parent.

It can hardly in this experiment be contended, that the ſkylark miſtook them for her own neſtlings, be-

* I could alſo wiſh that the following experiment was tried. When a hedge-ſparrow hath laid all her eggs, a ſingle one of any other bird, as large as a cuckow, might be introduced, after which if either the neſt was deſerted, or the egg too large to be hatched, it would afford a ſtrong preſumption againſt this prevailing opinion. I muſt here alſo take notice, that Mr. Hunter, F. R. S. who hath diſſected hen cuckows, informs me that they are not incapacitated from hatching their eggs, as hath been ſuppoſed by ſome ornithologiſts.

† I am perſuaded that a cuckow is oftener an orphan, than any other neſtling, becauſe, from the curioſity which prevails with regard to this bird, the parents are eternally ſhot.

cauſe.

cause they differed greatly, not only in number and size, but in their habits, for nightingales and wrens perch, which a skylark is almost incapable of, though, by great assiduity, she at last taught herself the proper equilibre of the body.

I have likewise been witness of the following experiment: two robins hatched five young ones in a breeding cage, to which five others were added, and the old birds brought up the whole number, making no distinction between them.

The Aëdologie also mentions (which is a very sensible treatise on the nightingale *) that nestlings of all sorts may be reared in the same manner, by introducing them to a caged bird, which is supplied with the proper food.

Not only old birds, however, attend to this cry of distress from nestlings, but young ones also which are able to shift for themselves.

I have seen a chicken, not above two months old, take as much care of younger chickens, as the parent would have shewn to them which they had lost, not only by scratching to procure them food, but by covering them with her wings; and I have little doubt but that she would have done the same by young ducks.

I have likewise been witness of nestling thrushes of a later brood, being fed by a young bird which was hatched earlier, and which indeed rather overcrammed the orphans instrusted to her care; if the bird however erred in judgement, she was certainly not deficient in tenderness, which I am persuaded she would have equally extended to a nestling cuckow.

* Paris, 1751, or 1771.

XXII. ΚΟΣ-

Received February 13, 1772.

XXII. ΚΟΣΚΙΝΟΝ ΕΡΑΤΟΣΘΕΝΟΥΣ.

OR,

The Sieve of Eratosthenes.

Being an account of his method of finding all the Prime Numbers, by the Rev. Samuel Horsley, *F. R. S.*

Read May 7, 1772.

A Prime number is such a one, as hath no intregral divisor but unity.

A number, which hath any other integral divisor, is Composite.

Two or more numbers, which have no common integral divisor, besides unity, are said to be Prime with respect to one another.

Two or more numbers, which have any common integral divisor besides unity, are said to be Composite with respect to one another.

The distinction of numbers into Prime and Composite, is so generally understood, that I suppose it is needless to enlarge upon it.

To determine, whether several numbers proposed be Prime or Composite *with respect to one another*, is an easy Problem. The solution of it is given by Euclid, in the three first propositions of the 7th book

book of the Elements, and is to be found in many common treatiſes of Arithmetic and Algebra. But to determine, concerning any number propoſed, whether it be *abſolutely* Prime or Compoſite, is a Problem of much greater difficulty. It ſeems indeed incapable of a direct ſolution, by any general method; becauſe the ſucceſſive formation of the prime numbers doth not ſeem reducible to any general law. And for the ſame reaſon, no direct method hath hitherto been hit upon, for conſtructing a Table of all the prime numbers to any given limit. Eratoſthenes, whoſe ſkill in every branch of the philoſophy and literature of his times, rendered his name ſo famous among the Sages of the Alexandrian School, was the inventor of an indirect method, by which ſuch a table might be conſtructed, and carried to a great length, in a ſhort time, and with little labour. This extraordinary and uſeful invention is at preſent, I believe, little, if at all, known; being deſcribed only by two writers, who are ſeldom read, and by them but obſcurely; by Nicomachus Geraſinus, a ſhallow writer of the 3d or 4th century, who ſeems to have been led into mathematical ſpeculations, not ſo much by any genius for them, as by a fondneſs for the myſteries of the Pythagorean and Platonic philoſophy; and by Boethius, whoſe treatiſe upon numbers is but an abridgment of the wretched performance of Nichomachus*. I flatter myſelf therefore, that a ſuccinct account of it will not be unacceptable to this learned Society.

* There are more pieces than one of this Nichomachus extant. That which I refer to is intitled Εισαγωγη Αριθμητικη.

1 But

But before I enter expresſly upon the ſubject, I muſt take the liberty to animadvert upon a certain Table, which, among other pieces aſcribed to Eratoſthenes, is printed at the end of the beautiful edition of Aratus publiſhed at Oxford in the year 1672, and is adorned with the title of Κοσκινον Ερατοσθενους. It contains all the odd numbers from 3 to 113 incluſive, diſtributed in little cells, all the diviſors of every Compoſite number being placed over it, in its proper cell, and the Prime numbers are diſtinguiſhed, ſo far as the table goes, by having no diviſors placed over them. It hath probably been copied either from a Greek comment upon the Arithmetic of Nicomachus, preſerved among the manuſcripts of Mr. Selden in the Bodleian Library, in which, though the manuſcript is now ſo much decayed as to be in moſt places illegible, I find plain veſtiges of ſuch a table *, which might be more perfect 100 years ago, when the Oxford Aratus was publiſhed; or elſe, from another comment, tranſlated from a Greek manuſcript into Latin, and publiſhed in that language, by Camerarius, in which a table of the very ſame form occurs, extending from the number 3 to 109 incluſive. It may ſufficiently ſkreen the editor of Aratus from cenſure, that he had theſe authorities to publiſh this table as the Sieve of Eratoſthenes; eſpecially as they are in ſome meaſure ſupported by paſſages of Nicomachus himſelf. But the Sieve of Eratoſthenes was quite another thing.

* This manuſcript ſeems to have contained the text of Nicomachus with Scholia in the margin. But the table evidently belongs to the Scholia, not to the text.

The Oxford editor hath annexed to his table, to explain the uſe of it, ſome detached paſſages, which he hath ſelected from the text of Nicomachus, and from a comment upon Nicomachus aſcribed to Joannes Grammaticus. In theſe paſſages the difference between Prime and Compoſite numbers is explained, in many words indeed, but not with the greateſt accuracy; and it is propoſed to frame a kind of Table of all the odd numbers, from 3 to any given limit, in which the Compoſite numbers ſhould be diſtinguiſhed by certain marks *. The Primes would conſequently be characteriſed, as far as the table ſhould be carried, by being unmarked. But, upon what principles, or by what rule, ſuch a table is to be conſtructed, is not at all explained. It is obvious that, in order to *mark* the Compoſite numbers, it is neceſſary to know which are ſuch. And, without ſome rule to diſtinguiſh which numbers are Prime, and which are Compoſite, independent of any table in which they ſhall be diſtinguiſhed by marks, it is impoſſible to judge, whether the table be true, as far as it goes, or to extend it, if requiſite, to a further limit. Now it was the Rule by which the Prime numbers and the Compoſite might be diſtinguiſhed, not a Table conſtructed we know not how, that was the invention of Eratoſthenes, to which from its uſe, as well as from the nature of the operation, which

* Nicomachus and Joannes Grammaticus propoſe that theſe marks ſhould be ſuch, as ſhould not only diſtinguiſh the compoſite numbers, but likewiſe ſerve to expreſs all the diviſors of every ſuch number. It will be ſhewn, in a proper place, that this was no part of the original contrivance of the Sieve.

5 proceeds

proceeds (as will be ſhewn) by a gradual extermination of the compoſite numbers from the arithmetical ſeries 3. 5. 7. 9. 11. &c. infinitely continued, its author gave the name of the Sieve. I have thought it neceſſary to premiſe theſe remarks, to remove a prejudice, which I apprehend many may have conceived, as this beautiful and valuable edition of Aratus is in every ones hands, that this ill-contrived table, the uſeleſs work of ſome monk in a barbarous age, was the whole of the invention of the great Eratoſthenes, and in juſtice to myſelf, that I might not be ſuſpected of attempting to reap another's harveſt.

I now proceed, to give a true account of this excellent invention; which, for its uſefulneſs, as well as for its ſimplicity, I cannot but conſider as one of the moſt precious remnants of Ancient Arithmetic. I ſhall venture to repreſent it according to my own ideas, not obliging myſelf to conform, in every particular, to the account of Nicomachus, which I am perſuaded is in many circumſtances erroneous. In ſtating the principles upon which the Operation of the Sieve was founded, he hath added obſervations upon certain relations of the odd numbers to one another, which are certainly his own, becauſe they are of no importance in themſelves, and are quite foreign to the purpoſe. Every thing of this kind I omit: and having ſtated what I take to have been the genuine Theory of Eratoſthenes's method, cleared from the adulterations of Nicomachus, I deduce from it an operation of great ſimplicity, which ſolves the Problem in queſtion with wonderful eaſe, and which, because it is the most simple that the theory seems to afford, I scruple not to adopt as the original Operation of the Sieve, though nothing like it is to be found in Nicomachus; though, on the contrary, Nichomachus, and all his Commentators, would suggest an operation very different from it, and far more laborious. For the satisfaction of the curious and the learned, I have annexed a copy of so much of Nicomachus's treatise, as relates to this subject, with such corrections of the text, as it stands in the edition of Wichelius, printed at Paris ann. 1538, as the sense hath suggested to me, or I have thought proper to adopt, upon the authority of a manuscript preserved among those of Archbishop Laud, in the Bodleian Library; which, in this part, I have carefully collated. By comparing this with the account which I subjoin, every one will be able to judge how far I have done justice to the invention I have undertaken to explain.

PROBLEM.

To find all the Prime Numbers.

The number 2 is a Prime number; but, except 2, no even number is Prime, because every even number, except 2, is divisible by 2, and is therefore Composite. Hence it follows, that all the Prime numbers, except the number 2, are included in the series of the odd numbers, in their natural order, infinitely extended; that is, in the series

3. 5. 7. 9. 11. 13. 15. 17. 19. 21. 23. 25. 27. 29. 31. 33. 35. 37. 39. 41. 43. 45. 47. 49. 51. &c.

Every

Every number which is not Prime, is a multiple of ſome Prime number, as Euclid hath demonſtrated (Element. 7. prop. 33.) Therefore the foregoing ſeries conſiſts of the Prime numbers, and of multiples of the Primes. And the multiples, of every number in the ſeries, follow at regular diſtances; by attending to which circumſtance, all the multiples, that is, all the Compoſite numbers, may be eaſily diſtinguiſhed and exterminated.

I ſay, the multiples of all numbers, in the foregoing ſeries, follow at regular diſtances.

For between 3 and its firſt multiple in the ſeries (9) two numbers intervene, which are not multiples of 3. Between 9 and the next multiple of 3 (15) two numbers likewiſe intervene, which are not multiples of 3. Again between 15 and the next multiple of 3 (21) two numbers intervene, which are not multiples of 3; and ſo on. Again, between 5 and its firſt multiple (15) four numbers intervene, which are not multiples of 5. And between 15 and the next multiple of 5 (25) four numbers intervene which are not multiples of 5; and ſo on. In like manner, between every pair of the multiples of 7, as they ſtand in their natural order in the ſeries, 6 numbers intervene which, are not multiples of 7. Univerſally, between every two multiples of any number *n*, as they ſtand in their natural order in the ſeries, $\overline{n-1}$ numbers intervene, which are not multiples of *n*.

Hence may be derived an Operation for exterminating the Compoſite numbers, which I take to have been the Operation of the Sieve, and is as follows.

The Operation of the Sieve.

Count all the terms of the ſeries following the number 3, by threes, and expunge every third number. Thus all the multiples of 3 are expunged. The firſt uncancelled number that appears in the ſeries, after 3, is 5. Expunge the ſquare of 5. Count all the terms of the ſeries, which follow the ſquare of 5, by fives, and expunge every fifth number, if not expunged before. Thus all the multiples of five are expunged, which were not at firſt expunged, among the multiples of 3. The next uncancelled number to 5 is 7. Expunge the ſquare of 7. Count all the terms of the ſeries following the ſquare of 7, by ſevens, and expunge every ſeventh number, if not expunged before. Thus all the multiples of 7 are expunged, which were not before expunged among the multiples of 3 or 5. The next uncancelled number which is now to be found in the ſeries, after 7, is 11. Expunge the ſquare of 11. Count all the terms of the ſeries, which follow the ſquare

---

3. 5. 7. ~~9~~. 11. 13. ~~15~~. 17. 19. ~~21~~. 23. ~~25~~. ~~27~~. 29. 31.
~~33~~. ~~35~~. 37. ~~39~~. 41. 43. ~~45~~. 47. ~~49~~. ~~51~~. 53. ~~55~~. ~~57~~. 59.
61. ~~63~~. ~~65~~. 67. ~~69~~. 71. 73. ~~75~~. ~~77~~. 79. ~~81~~. 83. ~~85~~. ~~87~~.
89. ~~91~~. ~~93~~. ~~95~~. 97. ~~99~~. 101. 103. ~~105~~. 107. 109. ~~111~~. 113.
~~115~~. ~~117~~. ~~119~~. ~~121~~. ~~123~~. ~~125~~. 127. ~~129~~. 131. ~~133~~. ~~135~~.
137. 139. ~~141~~. ~~143~~. ~~145~~. ~~147~~. 149. 151. ~~153~~. ~~155~~. 157.

of

of 11, by elevens, and expunge every eleventh number, if not expunged before. Thus all the multiples of 11 are expunged, which were not before expunged among the multiples of 3, 5, and 7. Continue these expunctions, till the first uncancelled number that appears, next to that whose multiples have been last expunged, is such, that its square is greater than the last and greatest number to which the series is extended. The numbers which then remain uncancelled are all the Prime numbers, except the number 2, which occur in the natural progression of number from 1 to the limit of the series. By the limit of the series I mean the last and greatest number to which it is thought proper to extend it.

Thus the prime numbers are found to any given limit.

Nicomachus proposes to make such marks over the Composite numbers, as should shew all the divisors of each. From this circumstance, and from the repeated intimations both of Nicomachus, and his commentator Joannes Grammaticus *, one would be led to imagine, that the Sieve of Eratosthenes was something more than its name imports, a method of sifting out the Prime numbers from the indiscriminate mass of all numbers Prime and Composite, and that, in some way or other, it exhibited all the divisors of every Composite number, and likewise shewed whether two or

* The Comment of Joannes Grammaticus is extant in manuscript in the Savilian Library at Oxford, to which I have frequent access, by the favour of the Reverend and Learned Mr. Hornsby, the Savilian Posessor of Astronomy.

more

more Composite numbers were Prime or Composite with respect to each other. I have many reasons to think, that this was not the case. I shall as briefly as possible point out some of the chief, for the matter is not so important, as to justify my troubling the Society with a minute detail of them. First then, in the natural series of odd numbers, 3. 5. 7. &c. every number is a divisor of some succeeding number. Therefore if we are to have marks for all the different divisors of every Composite number, we must have a different mark for every odd number. Therefore we must have as many marks, or systems of marks, as numbers; and I do not see, that it would be possible, to find any more compendious marks, than the common numeral characters. This being the case, it would be impracticable to carry such a table as Nicomachus proposes, and his commentators have sketched, to a sufficient length to be of use, on account of the multiplicity of the divisors of many numbers, and the confusion which this circumstance would create *. It is hardly to be supposed, that Eratosthenes could overlook this obvious difficulty, though Nicomachus hath not attended to it. Eratosthenes therefore could not intend the construction of such a table.

In the next place, such a table not being had, Eratosthenes could not but perceive, that, the determining whether two or more numbers be Prime or Composite with respect to one another, is in all cases to be done more easily, by the direct method given by Euclid, than by

* The number 3465 hath no less than 22 different divisors.

the

the method of the Sieve. And he could not mean, to apply this method to a problem, to which another was better adapted.

Laſtly, Eratoſthenes could not mean, that the method of the Sieve ſhould be applied to the finding of all the poſſible diviſors of any Compoſite number propoſed, becauſe he could not be unacquainted with a more ready way of doing this, founded upon two obvious Theorems, which could not be unknown to him.

The Theorems I mean are theſe.

1ſt. If two Prime numbers multiply each other, the number produced hath no diviſors but the two prime factors.

2d. If a Prime number multiply a Compoſite number, and likewiſe multiply all the diviſors of that compoſite ſeverally, the numbers produced by the multiplications of theſe diviſors will be diviſors of the number produced by the firſt multiplication: And the number produced by the firſt multiplication will have no diviſors, but the two factors, the diviſors of the Compoſite factor, and the numbers made by the multiplication of theſe diviſors by the Prime factor ſeverally.

The method of finding all the diviſors of any Compoſite number, delivered by Sir Iſaac Newton in the Arithmetica Univerſalis, and by Mr. Maclaurin in his Treatiſe of Algebra, may be deduced from theſe propoſitions, as every mathematician will eaſily perceive. This method requires indeed that the leaſt prime diviſor ſhould be previouſly found; and, if the leaſt prime diviſor ſhould happen to be a large number, as it is not aſſignable by any general method, the investigation of it by repeated tentations may be very tedious. A table therefore of the odd numbers *, in which the Composite numbers should each have its least Prime divisor written over it, would be very useful. But Nichomachus's project of framing a table in which each Composite number should have *all* its divisors written over it, is ridiculous and absurd, on account of the insuperable difficulties which would attend the execution of it.

Feb. 7, 1772.

S. Horsley.

* A table of the odd numbers would be sufficient: for the number 2 is the least prime divisor of every even number; and it is easy, even in the largest numbers, to try whether they are divisible by 2. In our method of notation, this may always be known, by observing the last figure in the expression of the number proposed.

EXCERPTA QUÆDAM
EX
Arithmeticâ Nicomachi
Ad Cribrum Eratoſthenis pertinentia.

Ἡ δὲ τούτων γένεσις (*a*), ὑπὸ Ἐρατοσθένους, καλεῖται Κόσκινον· ἐπειδὴ ἀναπεφυρμένους τοὺς περισσοὺς λαβόντες καὶ ἀδιακρίτους, ἐξ αὐτῶν [τὰ διαφέροντα ἀλλήλων εἴδη] (*b*) ταύτῃ τῇ τῆς γενέσεως (*c*) μεθόδῳ διαχωρίζομεν, ὡς δι᾽ ὀργάνου ἢ κοσκίνου τινός· καὶ ἰδίᾳ μὲν τοὺς πρώτους καὶ ἀσυνθέτους, χωρὶς δὲ τοὺς μίκτους εὑρίσκομεν. Ἔστι δὲ ὁ τρόπος τοῦ Κοσκίνου τοιοῦτος. Ἐκθέμενος τοὺς ἀπὸ τριάδος πάντας ἐφεξῆς περισσοὺς, ὡς δυνατὸν μάλιστα ἐπὶ μήκιστον στίχον, ἀρξάμενος ἀπὸ τοῦ πρώτου, ἐπισκοπῶ τίνας οἷός τε ἔσται μετρεῖν ἕκαστος· καὶ εὑρίσκω δυνατὸν ὄντα τὸν πρῶτον, ἤτοι τὸν γ̄, τοὺς δύο μέσους διαλείποντας (*d*) μετρεῖν, μέχρις οὗ προχωρεῖν ἐθέλωμεν (*e*). οὐχ ὡς ἔτυχε δὲ, καὶ εἰκῇ, μετροῦντα, ἀλλὰ τὸν μὲν πρώτως αὐτῶν κείμενον, τοῦτ᾽ ἔστι τὸν ἀφ᾽ ἑαυτοῦ τοὺς δύο μέσους διαλεί-

(*a*) Mallem εὕρεσις, etſi, ne quid diſſimulem, lectioni receptæ adſtipulatur Boethii interpretatio.

(*b*) Voces uncis incluſas conjecturâ ſupplevi; quin et ſequentium ordinem paululum immutavi, pro τῇ γενέσεως μεθόδῳ ταύτῃ, ſcribendo ταύτῃ τῇ κ. τ. λ.

(*c*) Vocem γενέσεως hic loci retinendam cenſeo. Locum integrum ſic interpretor. "Suam horum indaginem Eratoſthenes, Cribrum vocavit. Propterea quod imparibus univerſis, nullo generum diſcrimine, in medio collocatis, ipſam procreationem continuam, quo tradidit ille modo, inſequendo [id eſt, procreationis continuæ, Eratoſthenis modo, exploratâ lege] ſpecies diverſas ſeorſim ſiſtimus, cribro tanquam ſeparatas."

(*d*) Cod. MS. habet διαλείποντα. Wechelius παραλείποντα.

(*e*) Ex Cod. MS. pro ἐθέλομεν.

πονlα (*f*), καlὰ τὴν τȣ̃ πρωlίςȣ ἐν τῷ ςίχῳ κειμένȣ ποσότηlα μεlρήσει· τȣ̃τ' ἔςι καlὰ τὴν ἑαυlȣ̃, τρὶς γάρ· τὸν δ' ἀπ' ἐκείνȣ δύο διαλείπονlα, καlὰ τὴν τȣ δευlέρȣ τεταγμένȣ, πενlάκις γάρ· τὸν δὲ περαιlέρω πάλιν δύο διαλείπονlα, καlὰ τὴν τȣ̃ τρίτȣ τεlαγμένȣ, ἑπlάκις γάρ· τὸν δὲ ἔτι περαιlέρω ὑπὲρ δύο κείμενον, καlὰ τὴν τȣ̃ τεlάρlȣ τεlαγμένȣ, ἐννεάκις γάρ· καὶ ἐπ' ἄπειρον τῷ αὐτῷ τρόπῳ. Εἶτα μεlὰ τȣ̃τον, ἀπ' ἄλλης ἀρχῆς, ἐπὶ τὸν δεύτερον ἐλθὼν, σκοπῶ τίνας οἷός τε ἔςι μεlρεῖν· καὶ εὑρίσκω πάνlας τὰς τέσαρας (*g*) διαλείπονlας· ἀλλὰ τὸν μὲν πρῶτον, καlὰ τὴν ἐν τῷ ςίχῳ πρωlίςȣ τεlαγμένȣ ποσότηlα· τρὶς γάρ. τὸν δὲ δεύτερον, καlὰ τὴν τȣ̃ δευlέρȣ· πενlάκις γάρ· τὸν δὲ τρίτον, καlὰ τὴν τȣ̃ τρίτȣ· ἑπlάκις γάρ· καὶ τȣ̃το ἐφεξῆς ἀεί. Πάλιν δὲ ἄνωθεν, ὁ τρίτος, ὁ ζ̄, τὸ μετρεῖν* παραλαβὼν, μεlρήσει τὰς ἓξ διαλείπονlας· ἀλλὰ τὸν μὲν πρώτιςον, καlὰ τὴν τȣ̃ γ̄ (*h*) ποσότηlα, πρώτȣ κειμένȣ· τὸν δὲ δεύτερον καlὰ τὴν τȣ̃ ε̄· δευlερόlαγὴς γὰρ ȣ̃τος (*i*). τὸν δὲ τρίτον, καlὰ τὴν τȣ̃ ζ̄, τρίτην γὰρ ἔχει (*k*) ȣ̃τος τάξιν ἐν τῷ ςίχῳ. καὶ, καlὰ τὴν αὐτὴν ἀναλογίαν, δι' ὅλον (*l*) ἀπαραποδίςως (*m*) προχωρήσει σοι τȣ̃το, ὥςε τὸ μὲν μεlρεῖν διαδέξονlαι, καlὰ τὴν ἐν τῷ ςίχῳ αὐτῶν ἐγκειμένην τάξιν· τὸ δὲ πόσȣς διαλείποντας,

(*f*) Locum in Editione Wechelii corruptum, in Cod. MS. mutilum & turbatum, conjecturâ, prout potui, ſanatum dedi. Editio Wechelii habet τὸν τȣ̀ς δύο μέσȣς ὑπερβαίνονlα. Codex MS. τὸν δύο. τελέσι τὸν τρία.

(*g*) Conjecturâ, pro τέlρασι.

(*h*) Litera numeralem γ̄, conjecturâ poſui pro voce τρια.

(*i*) Reſtitui ex Cod. MS pro ὄνlος, quæ eſt Wechelii lectio.

(*k*) Particulam καὶ omiſi.

(*l*) Wechelium ſequor. Cod. MS. habet λογȣ, ſenſu, ut videtur, nullo.

(*m*) Ex Cod. MS. pro ἀπαρεμπόδιςον.

* Conjecturâ pro μέlρον.

καlὰ

κατὰ τὴν ἀπὸ δυάδος ἐπ' ἄπειρον εὔτακτον τῶν (n) ἀρτίων προκοπὴν, ἢ κατὰ τὴν τῆς χώρας διπλασίασιν καθ' ἣν ὁ μετρῶν τέτακται· τὸ δὲ ποσάκις, κατὰ τὴν τῶν ἀπὸ τριάδος περιττῶν εὔτακτον ἐπ' ἄπειρον (o) προχώρησιν (p). Ἐὰν οὖν σημείοις τισὶν ἐπισίξῃς τοὺς ἀριθμοὺς, εὑρήσεις τοὺς μεταλαμβάνοντας τὸ μετρεῖν, οὔτε ἅμα πάντας τὸν αὐτόν ποτε μετροῦντας, ἔστι δὲ ὅτε οὐδὲ δύο τὸν αὐτόν· οὔτε πάντας ἁπλῶς τοὺς ἐκκειμένους ὑποπίπτοντας μέτρῳ τινὶ αὐτῶν. ἀλλὰ τινὰς μὲν παντελῶς διαφεύγοντας τὸ μετρηθῆναι ὑφ' οὑτινοσοῦν· τινὰς δὲ ὑφ' ἑνὸς μόνου μετρουμένους· τινὰς δὲ ὑπὸ δυὸ, ἢ καὶ πλειόνων. Οἱ μὲν οὖν μηδαμῶς (q) μετρηθέντες, ἀλλὰ διαφυγόντες τοῦτο, πρῶτοι εἰσὶ καὶ ἀσύνθετοι, ὡς ὑπὸ κοσκίνου διακριθέντες. οἱ δὲ ὑφ' ἑνὸς μόνου μετρηθέντες, κατὰ τὴν ἑαυτοῦ (r) ποσότητα, ἓν μόνον μόριον ἑτερώνυμον ἕξουσι πρὸς τῷ παρωνύμῳ. οἱ δὲ ὑφ' ἑνὸς μὲν (s), ἑτέρου δὲ ποσότητι, καὶ μὴ τῇ ἑαυτοῦ, ἢ ὑπὸ δύο ὁμοῦ μετρηθέντες, πλείονα ἕξουσι τὰ ἑτερώνυμα μέρη πρὸς τῷ παρωνύμῳ. τοῦτοι οὖν ἔσονται

(n) Conjectura pro τὴν.

(o) Voces ἐπ' ἄπειρον ex Cod. MS. restitui.

(p) Nempe series numerorum imparium 3, 5, 7, 9, &c. infinite protensa, cum numeros impares universos contineat, imparis cujusvis multiplices omnes impares necessario complectitur. Esto igitur n numerus quilibet impar. In serie 3, 5, 7, &c. infinite protensâ, habes numeros omnes $\overline{n\times 3}$, $\overline{n\times 5}$, $\overline{n\times 7}$, $\overline{n\times 9}$, &c. Et cum seriei ea Lex sit & Conditio, ut naturali ordine numeri impares sequantur, & minor omnis numerus majorem præcedat, fieri nequit, quin multiplices numeri n eum inter se ordinem servent, ut minor quisque majorem præcedat. Primus igitur erit $\overline{n\times 3}$, secundus $\overline{n\times 5}$, tertius $\overline{n\times 7}$, & universim, $\overline{n\times m}$ eum habiturus est, inter multiplices, locum, quem numerus m in serie.

(q) Ex Cod. MS. vice οὐδαμῶς, quæ Wechelii lectio est.

(r) Conjecturâ pro ἑαυτῶν.

(s) Particulam μὲν ex Cod. MS restitui.

δεύτεροι καὶ σύνθετοι. Τὸ δὲ τρίτον μέρος, τὸ κοινὸν ἀμφοτέρων, ὁ καθ' ἑαυτὸ μὲν δεύτερον καὶ σύνθετον, πρὸς ἄλλο δὲ πρῶτον καὶ ἀσύνθετον, ἔσονται ἀποτελούμενοι ἀριθμοί, κατὰ τὴν ἑαυτοῦ ποσότητα πρώτου καὶ ἀσυνθέτου μετρήσαντος τινός, εἴτις [τούτῳ τῷ τρόπῳ] (*t*) γινόμενος, συγκρίνοιτο πρὸς ἄλλον ὡσαύτως τὴν γένεσιν ἔχοντα. ὥσπερ ὁ θ̄, ἐγένετο γὰρ ἐκ τοῦ γ̄ (*u*) κατὰ τὴν ἑαυτοῦ ποσότητα μετρήσαντος· τρὶς γάρ· εἰ συγκρίνοιτο πρὸς τὸν κε̄· ἐγένετο γὰρ καὶ οὗτος (*x*) ἐκ τοῦ ε̄, κατὰ τὴν ἑαυτοῦ ποσότητα μετρήσαντος· πεντάκις γάρ· κοινὸν μέτρον τούτοις οὐκ ἔσται, εἰ μὴ μόνη ἡ Μονάς.

(*t*) Voces τούτῳ τῷ τρόπῳ conjecturâ ſupplevi.

(*u*) Literam numeralem γ̄ pro voce τρίτου quæ apud Wechelium legitur, ex Cod. MS reſtitui.

(*x*) Voces γὰρ καὶ οὗτος ex Cod. MS. reſtitui.

Ex

Ex Arithmeticâ Boethii.

Lib. I. c. xvii.

GENERATIO autem ipsorum atque ortus hujusmodi investigatione colligitur, quam scilicet Eratosthenes Cribrum nominabat; quod cunctis imparibus in medio collocatis, per eam, quamādituri sumus, artem, qui primi, quive secundi, quique tertii generis videantur esse distinguitur. Disponantur enim a ternario numero cuncti in ordinem impares, in quamlibet longissimam porrectionem 3. 5. 7. 9. 11. 13. 15. 17. 19. 21. 23. 25. 27. 29. 31. 33. 35. 37. 39. 41. 43. 45. 47. 49. His igitur ita dispositis, considerandum, primus numerus quem eorum, qui sunt in ordine positi, primum metiri possit: sed, duobus præteritis, illum, qui post eos est positus, mox metitur: et, si post eundem ipsum quem mensus est, alii duo transmissi sunt, illum, qui post duos est, rursus metitur: et, eodem modo si duos quis reliquerit, post eos qui est, a primo numero metiendus est; eodemque modo, relictis semper duobus, a primo, in infinitum pergentes metientur. Sed id non vulgo neque confuse. Nam primus numerus illum, qui est post duos secundum se locatos, per suam quantitatem metitur: ternarius enim numerus ter [a] 9 metitur. Si autem post novenarium duos reliquero, qui mihi post illos incurre-

[a] Conjecturâ pro *tertio*.

rit, a primo metiendus est, per secundi imparis quantitatem; id est, per quinarium: nam si post 9 duos relinquam, id est 11 & 13, ternarius numerus 15 metietur, per secundi numeri quantitatem, id est, per quinarii; quoniam numerus ternarius 15 quinquies metitur. Rursus, si a quindenario inchoans duos intermisero, qui posterior positus est, ejus primus numerus mensura est, per tertii imparis pluralitatem: nam si post 15 intermisero 17 & 19, incurrit 21, quem ternarius numerus secundum septenarium metitur; 21 enim numeri ternarius septima pars est: atque hoc in infinitum faciens, reperio primum numerum, si binos intermisero, omnes sequentes post se metiri, secundum quantitatem positorum ordine imparium numerorum. Si vero quinarius numerus, qui in secundo loco est constitutus, velit [b] quis, cujus prima ac deinceps sit mensura, invenire, transmissis quatuor imparibus, quintus ei quem metiri possit, occurrit. Intermittantur enim quatuor impares, id est, 7 & 9, & 11 & 13, post hos est quintus decimus quem quinarius metitur, secundum primi scilicet quantitatem, id est, ternarii; quinque enim 15 ter[c] metiuntur: ac deinceps, si quatuor intermittat, eum qui post illos locatus est, secundus, id est, quinarius, sui quantitate metitur: nam post quindecim intermissis 17 & 19, & 21 & 23, post eos 25 reperio, quos quinarius scilicet numerus suâ pluralitate metitur; quinquies enim quinario multiplicato, 25 succrescunt; si vero post hunc quilibet quatuor intermittat, eâdem ordinis servatâ

[b] Conjecturâ pro *vel.*
[c] Conjecturâ pro *tertio.*

constantiâ,

conſtantiâ, qui eos ſequitur, ſecundum tertii, id eſt, ſeptenarii numeri ſummam, a quinario metitur: atque hæc eſt infinita proceſſio. Si vero tertius numerus quem metiri poſſit exquiritur, ſex in medio relinquentur; & quem ſeptimum ordo monſtraverit, hic per primi numeri, id eſt, ternarii quantitatem metiendus eſt: et poſt illum, ſex aliis interpoſitis, quem poſt eos numeri ſeries dabit, per quinarium, id eſt, per ſecundum, tertii eum menſura percurret: ſi vero alios rurſus ſex in medio quis relinquat, ille, qui ſequitur, per ſeptenarium ab eodem ſeptenario metiendus eſt; id eſt, per tertii quantitatem; atque hic uſque in extremum ratus ordo progreditur. Suſcipient ergo metiendi viciſſitudinem, quemadmodum ſunt in ordine naturaliter impares conſtituti: metientur autem, ſi per pares numeros, a binario inchoantes, poſitos inter ſe impares, ratâ intermiſſione, tranſiliant; ut primus duos, ſecundus quatuor, tertius ſex, quartus octo, quintus decem[d]: vel ſi locos ſuos conduplicent, & ſecundum duplicationem terminos intermittant; ut ternarius, qui primus eſt numerus, & Unus, omnis enim primus Unus eſt, bis locum ſuum multiplicet, faciatque bis unum; qui cum duo ſint, primus duos medios tranſeat. Rurſus ſecundus, id eſt, quinarius, ſi locum ſuum multiplicet, 4 explicabitur: hic quoque quatuor [e] intermittat. Item ſi ſeptenarius, qui tertius eſt, locum ſuum duplicet, ſex creabit; bis enim 3 ſenarium jungunt: hic ergo in ordidine [f] ſex relinquat. Quartus quoque, ſi locum

[d] Conjecturâ reſtitui pro 12.

[e] Conjecturâ pro 4.

[f] Conjecturâ pro *ordinem*.

ſuum duplicet, 8 ſuccreſcent; ille quoque octo tranſiliat: atque hoc quidem in cæteris perſpiciendum. Modum autem menſionis, ſecundum ordinem collocatorum, ipſa ſeries dabit. Nam primus primum quem numerat, ſecundum primum numerat [g]; id eſt, ſecundum ſe; & ſecundum primus quem numerat, per ſecundum numerat [g], & tertium per tertium, & quartum item per quartum. Cum autem ſecundus menſionem [h] ſuſceperit, primum quem numerat ſecundum primum metitur; ſecundum vero quem numerat per ſe, id eſt, per ſecundum; & tertium per tertium: & in cæteris eâdem ſimilitudine menſura conſtabit. Illos [i] ergo ſi reſpicias, vel qui alios menſi ſunt, vel qui ipſi ab aliis metiuntur, invenies omnium ſimul communem menſuram eſſe non poſſe, neque ut omnes quemquam alium ſimul numerent; quoſdam autem ex his ab alio poſſe metiri, ita ut ab uno tantum numerentur [k]; alios vero, ut etiam a pluribus; quoſdem autem, ut præter Unitatem eorum nulla menſura ſit. Qui ergo nullam menſuram præter Unitatem recipiunt, hos Primos & Incom-

[f] Conjecturâ pro 8.

[g] Pro *numerat* mallem in utroque loco, *metitur*, ut aliud ſit *numerare*, aliud *metiri*, & ſenſus ſit, "That which the firſt "number [of the Series] *counts* the firſt [of its multiples], it "*meaſures* by the firſt [of the Series], i. e. by itſelf. That "which it *counts* the ſecond [of its multiples], it *meaſures* by "the ſecond [number in the Series]." Sic enim infra legimus de Numero ordine ſecundo, "primum quem *numerat* ſeundum "primum *metitur*."

[h] Conjecturâ, pro *manſionem*.

[i] Conjectura, pro *alios*.

[k] Ang. "But ſo as to be counted in among the multiples of "one number only."

 poſitos

positos judicamus; qui vero aliquam mensuram præter Unitatem, vel alienigenæ partis vocabulum sortiuntur, eos pronunciemus Secundos atque Compositos. Tertium vero illud genus, per se Secundi & Compositi, Primi vero & Imcompositi ad alterutrum comparati, hâc inquisitor ratione reperiet. Si enim quoslibet primos [l] numeros, secundum suam in semetipsos multiplices quantitatem, qui procreantur, ad alterutrum comparati, nullâ mensurâ communione junguntur: 3 [m] enim & 5, si multiplices, 3 ter [n] 9 faciunt, & quinquies 5 reddunt 25. His igitur nulla est cognatio communis mensuræ. Rursus 5 & 7 quos procreant, si compares, hi quoque incommensurabiles erunt: quinquies enim 5, ut dictum est, 25, septies 7 faciunt 49; quorum mensura nulla communis est, nisi forte omnium horum procreatrix & mater Unitas [o].

[l] Conjectura pro *illos*.
[m] Conjecturâ, pro *tres*. [n] Conjecturâ pro *tres tertio*.
[o] Sed cave credas, Lector, numeros inter se primos nullos dari præter Primorum Quadratos.

XXIII. *A Letter from Mr.* Chriſtopher Gullet *to* Matthew Maty, *M. D. Sec. R. S. on the Effects of Elder, in preſerving Growing Plants from Inſects and Flies.*

Taviſtock (Devon) Auguſt 11, 1771.

SIR,

Read May 14, 1771.

I SHOULD not preſume to trouble you as a member of the Royal Society with the following letter, did not the ſubject ſeem to promiſe to be of great public utility. It relates to the effects of Elder;

Sambucus fructu in umbella nigro.

1ſt. In preſerving cabbage plants from being eaten or damaged by caterpillers.

2d. In preventing blights, and their effects on fruit and other trees.

3d. In the preſervation of crops of wheat from the yellows, and other deſtructive inſects.

4th. Alſo in ſaving crops of turnips from the fly, &c. &c.

1ſt, I was led to my firſt experiments, by conſidering how diſagreeable and offenſive to our olfactory nerves the effluvia emitted by a bruſh of green elder

elder leaves are, and from thence, reaſoning how much more ſo they muſt be to thoſe of a butterfly, whom I conſidered as being as much ſuperior to us in delicacy as inferior in ſize. Accordingly I took ſome twigs of young elder, and with them whipt the cabbage plants well, but ſo gently as not to hurt them, juſt as the butterflies firſt appeared; from which time, for theſe two ſummers, though the butterflies would hover and flutter round them like gnomes or ſylphs, yet I could never ſee one pitch, nor was there I believe a ſingle catterpiller blown, after the plants were ſo whipt; though an adjoining bed was infeſted as uſual.

2d. Reflecting on the effects abovementioned, and conſidering blights as chiefly and generally occaſioned by ſmall flies, and minute inſects, whoſe organs are proportionably finer than the former, I whipt the limbs of a wall plumb tree, as high as I could reach; the leaves of which were preſerved green, flouriſhing, and unhurt, while thoſe not ſix inches higher, and from thence upwards, were blighted, ſhrivelled up, and full of worms. Some of theſe laſt I afterwards reſtored by whipping with, and tying up, elder among them. It muſt be noted, that, this tree was in full bloſſom at the time of whipping, which was much too late, as it ſhould have been done once or twice before the bloſſom appeared. But I conclude from the whole, that if an infuſion of elder was made in a tub of water, ſo that the water might be ſtrongly impregnated therewith, and then ſprinkled over the tree, by a hand engine, once every week or fortnight, it would effectually

anſwer

anſwer every purpoſe that could be wiſhed, without any poſſible riſk of hurting the bloſſoms or fruit.

3d. What the farmers call the yellows in wheat, and which they conſider as a kind of mildew, is in fact, as I have no doubt but you well know, occaſioned by a ſmall yellow fly with blue wings, about the ſize of a gnat. This blows in the ear of the corn, and produces a worm, almoſt inviſible to the naked eye; but being ſeen through a pocket microſcope, it appears a large yellow maggot of the colour and gloſs of amber, and is ſo prolific that I laſt week diſtinctly counted 41 living yellow maggots or inſects, in the huſk of one ſingle grain of wheat, a number ſufficient to eat up and deſtroy the corn in a whole ear. I intended to have tryed the following experiment ſooner; but the dry hot weather bringing on the corn faſter than was expected, it was got and getting into fine bloſſoms ere I had an opportunity of ordering as I did; but however the next morning at daybreak, two ſervants took two buſhes of elder, and went one on each ſide of the ridge from end to end, and ſo back again, drawing the elder over the ears of corn of ſuch fields as were not too far advanced in bloſſoming. I conceived, that the diſagreeable effluvia of the elder would effectually prevent thoſe flies from pitching their tents in ſo noxious a ſituation; nor was I diſappointed, for I am firmly perſuaded that no flies pitched or blowed on the corn after it had been ſo ſtruck. But I had the mortification of obſerving the flies (the evening before it was ſtruck) already on the corn (ſix, ſeven or eight, on a ſingle ear) ſo that what damage hath accrued, was done before

before the operation took place; for, on examining it last week, I found the corn which had been struck pretty free of the yellows, very much more so than what was not struck. I have, therefore, no doubt but that, had the operation been performed sooner, the corn would have remained totally clear and untouched. If so, simple as the process is, I flatter myself, it bids fair to preserve fine crops of corn from destruction, as the small insects are the crops greatest enemy. One of those yellow flies laid at least eight or ten eggs of an oblong shape on my thumb, only while carrying by the wing across three or four ridges, as appeared on viewing it with a pocket microscope.

4th. Crops of turnips are frequently destroyed, when young, by being bitten by some insects, either flies or fleas; this I flatter myself may be effectually prevented, by having an elder bush spread so as to cover about the breadth of a ridge, and drawn once forward and backward by a man over the young turnips. I am confirmed in this idea, by having struck an elder bush over a bed of young collyflower plants, which had begun to be bitten, and would otherwise have been destroyed by those insects; but after that operation it remained untouched.

In support of my opinion, I beg leave to mention the following fact from very credible information, that about eight or nine years ago this county was so infested with cock chaffers or oakwebs, that in many parishes they eat every green thing, but elder; nor left a green leaf untouched besides elder bushes, which alone remained green and unhurt, amid the general devastation of so voracious a multitude. On

reflecting

reflecting on these several circumstances, a thought suggested itself to me, whether an elder, now esteemed noxious and offensive, may not be one day seen planted with, and entwisting its branches among, fruit trees, in order to preserve the fruit from destruction of insects: and whether the same means which produced these several effects, may not be extended to a great variety of other cases, in the preservation of the vegetable kingdom.

The dwarf elder (*ebulus*) I apprehend emits more offensive effluvia then common elder, therefore must be preferable to it in the several experiments.

On mentioning lately to Sir Richard W. Bampfylde, one of the representatives of this county, my observations on the corn crops, and the effects of the elder, &c. he persuaded me to publish them, which in some measure determined my taking this step, of transmitting them to a Society incorporated for promoting the knowledge of natural things, and useful experiments, in which they have so happily and amply succeded, to the unspeakable advantage and improvement both of the old and new world. I have the honour to subscribe myself,

SIR,

Your most obedient,

humble Servant,

Chr. Gullett.

XXIV. *A*

XXIV. *A Letter from* John Call, *Esq*; *to* Nevil Maſkelyne, *F. R. S. Aſtronomer Royal, containing a Sketch of the Signs of the Zodiac, found in a Pagoda, near Cape* Comorin *in* India.

S I R,

Read May 14, 1772.

AS a member of the Royal Society, and one whoſe ſtudy is particularly directed to the motions of the heavenly bodies, I think you the moſt proper perſon to whom I can ſend the incloſed ſketch [Tab. X.], which I drew with a pencil, as I lay on my back reſting myſelf during the heat of the day, in a journey from Madurah to Twinwelly, near Cape Comorin. And I ſend it to you rather in the original, as I then ſketched it off, than in any more complete form, leſt it ſhould thereby have more the appearance of compoſition, and leave not ſo ſtrong an impreſſion of antiquity, as it made on me when I diſcovered it.

After ſuch a diſcovery, I ſearched in my travels many other pagodas, or choultrys, for ſimilar carvings; but, to the beſt of my remembrance, never found but one more equally complete, which was on the ceiling of a temple, in the middle of a tank before the pagoda of Teppecolum, near Mindurah, of which tank and temple Mr. Ward, painter in Broad-ſtreet, near Carnaby-market, hath a drawing; but I have often met with the ſeveral parts in detached pieces.

From the correſpondence of the ſigns of the zodiac which we at preſent uſe, and which we had, I believe, from the Arabians or Egyptians, I am apt to think that they originally came from India, and were in uſe among the Bramins, when Zoroaſter and Pythagoras travelled thither, and conſequently adopted and uſed by thoſe travellers: and as theſe philoſophers are ſtill ſpoken of in India, under the names of Zerdhurſt and Pyttagore, I ſhould alſo hazard another idea, that the worſhip of the cow, which ſtill prevails in India, was tranſplanted from thence to Egypt. But this is only conjecture; and it may with almoſt equal probability be ſaid, that Zoroaſter or Pythagoras carried that worſhip to India.

However, I think there is an argument ſtill in favour of India for its antiquity, in point of civilization and cultivation of the arts and ſciences; for it is hardly in diſpute that all theſe improvements came from the eaſt to the weſt; and, if we may be allowed to draw any concluſions from the immenſe buildings now exiſting, and from the little of the inſcriptions, which can be interpreted on ſeveral of the choultrys and pagodas, I think it may ſafely be pronounced, that no part of the world has more marks of antiquity for arts, ſciences, and civilization,

tion, than the peninſula of India, from the Ganges to Cape Comorin; nor is there in the world a finer climate, or face of the country, nor a ſpot better inhabited, or filled with towns, temples, and villages, than this ſpace is throughout, if China and parts of Europe are excepted.

I think the carvings on ſome of the pagodas and choultrys, as well as the grandeur of the work, exceeds any thing executed now-a-days, not only for the delicacy of the chiſſel, but the expence of conſtruction, conſidering, in many inſtances, to what diſtances the component parts were carried, and to what heights raiſed. If Mr. Kittle the painter, now in India, ſhould have time and opportunity, after he hath made his fortune by portrait drawing, it would be a great addition to his reputation, and well worth his pains, to inveſtigate the nature of the Indian architecture and carving, by painting ſome of the moſt curious buildings, or parts of pagodas. The great obſtacle to aſcertaining dates, or hiſtorical events, is the loſs of the Sans-Skirrit language, and the confinement of it to the prieſthood. I ſhould have taken ſome pains to have collected many things; but the number of revolutions and occupations which happened always prevented me.

I alſo commit to your inſpection the * manuſcripts of Mr. Robins, which he gave me at his death;

* Theſe I communicated to the Royal Society, together with this letter; but being examined by myſelf, Mr. Raper, Mr. Cavendiſh, and Mr. Horſley, at the deſire of the Society, they were not found to contain any thing material more than has been already printed; excepting a treatiſe on military diſcipline: which, if it ſhould be thought of uſe, may be inſerted in the next edition of his works. N. M.

I believe most of them have been printed, but if there are any which have not, or that can amuse you or instruct others, you are welcome to use them as you please: I only wish they may contain any thing useful. While he lived, I pursued those studies; but, soon after his death, new scenes arose, and engaged me more in practical service, than allowed me time for theory, or experiments. I am, however, a constant well-wisher to the progress of arts and sciences, as well as study; and very much,

SIR,

Your obedient,

humble servant,

Jn° Call.

XXV. *An*

XXV. *An Account of the Flowing of the Tides in the* South Sea, *as observed on board His Majesty's Bark the* Endeavour, *by Lieut.* J. Cook, *Commander, in a Letter to* Nevil Maskelyne, *Astronomer Royal, and F. R. S.*

Mile-end, February 5, 1772.

Reverend Sir,

Read May 21, 1772.

I Here send you the few observations I made on the tides in the South Sea, to which I have only to add, that, from many circumstances and observations, I am fully convinced that the flood comes from the southward, or rather from the S. E. I am,

S I R,

Your most obedient,

humble servant,

J. Cook.

Names

Names of places where obferved.	Lat. South.		Long. Weft.		New and full Moon. High water.		Rife & fall.	
					H.	M.	F.	In.
Succefs Bay in Strait le Maire —	54	45	66	4	4	30	5	6
Lagoon Ifland — — —	18	47	139	28	0	30	. .	. .
Matavai Bay, Otaheita — —	17	29	149	30	0	30	0	11
Tolaga Bay, Eaft coaft of New Zealand	38	22	181	14	6	0	5	6
Mercury Bay, N. E. ditto — —	36	48	184	4	7	30	7	0
River Thames, ditto — — —	37	12	184	12	9	0	10	0
Bay of Iflands, ditto — — —	35	14	185	36	8	0	7	0
Queen Charlotte's Sound, Cook's Strait New Zealand — — —	41	0	184	45	9	30	7	6
Admiralty Bay, in ditto — —	41	45	185	12	10	0	7	0
Botany Bay, coaft of New South-Wales	34	0	208	37	8	0	4	6
Buftard Bay, ditto — — —	24	30	208	20	8	0	8	0
Thirfty Sound, ditto — — —	25	5	210	24	11	0	16	0
Endeavour River, ditto — —	15	26	214	48	9	30	9	0
Endeavour's Strait, which divides New Guinea from New Holland —	10	37	218	45	1	30	11	0

XXVI. *An*

XXVI. *An Account of a new Electrometer, contrived by Mr.* William Henly, *and of several Electrical Experiments made by him, in a Letter from Dr.* Prieftley, *F.R.S. to Dr.* Franklin, *F. R. S.*

Dear Sir,

Read May 28, 1772.

I THINK myfelf happy in an opportunity of giving you a fpecies of pleafure, which I know is peculiarly grateful to you as the father of modern electricity, by tranfmitting to you an account of fome very curious and valuable improvements in your favourite fcience. The author of them is Mr. Henly, in the Borough, who has favoured me with the communication of them, and has given me leave to requeft, that you would prefent them to the Royal Society.

In my hiftory of electricity, and elfewhere, I have mentioned a good electrometer, as one of the greateft defiderata among practical electricians, to meafure both the precife degree of the electrification of any body, and alfo the exact quantity of a charge before the explofion, with refpect to the fize of the electrified body, or the jar or battery with which it is connected; as well as to afcertain the moment of time, in which the electricity of a jar changes, when, without making an explofion, it is difcharged by

giving

giving it a quantity of the contrary electricity. All these purposes are answered, in the most complete manner, by an electrometer of this gentleman's contrivance, a drawing of which I send you along with the following description.

The whole instrument is made of ivory or wood, [Tab. XI.] (*a*) is an exceeding light rod, with a cork ball at the extremity, made to turn upon the center of a semicircle (*b*), and so as always to keep pretty near the limb of it, which is graduated: (*c*) is the stem that supports it, and may either be fixed to the prime conductor, or be let into the brass knob of a jar or battery, or set in a stand, to support itself.

The moment that this little apparatus is electrified, the rod (*a*) is repelled by the stem (*c*), and consequently begins to move along the graduated edge of the semicircle (*b*); so as to mark with the utmost exactness, the degree in which the prime conductor, &c. is electrified, or the height to which the charge of any jar or battery is advanced; and as the materials of which this little instrument is made are very imperfect conductors, it will continue in contact with any electrified body, or charged jar, without dissipating any of the electricity.

If it should be found, by trial in the dark, that any part of this instrument contributes to the dissipation of the electric matter, (which, when the electrification was very strong, I once observed mine to do) it should be baked * a little, which will presently prevent it. If it is heated too much, it will not receive electricity readily enough; and then the motion of the index will not correspond with sufficient

* Warmed a little, to dry off the damps, particularly from the index.

exactness,

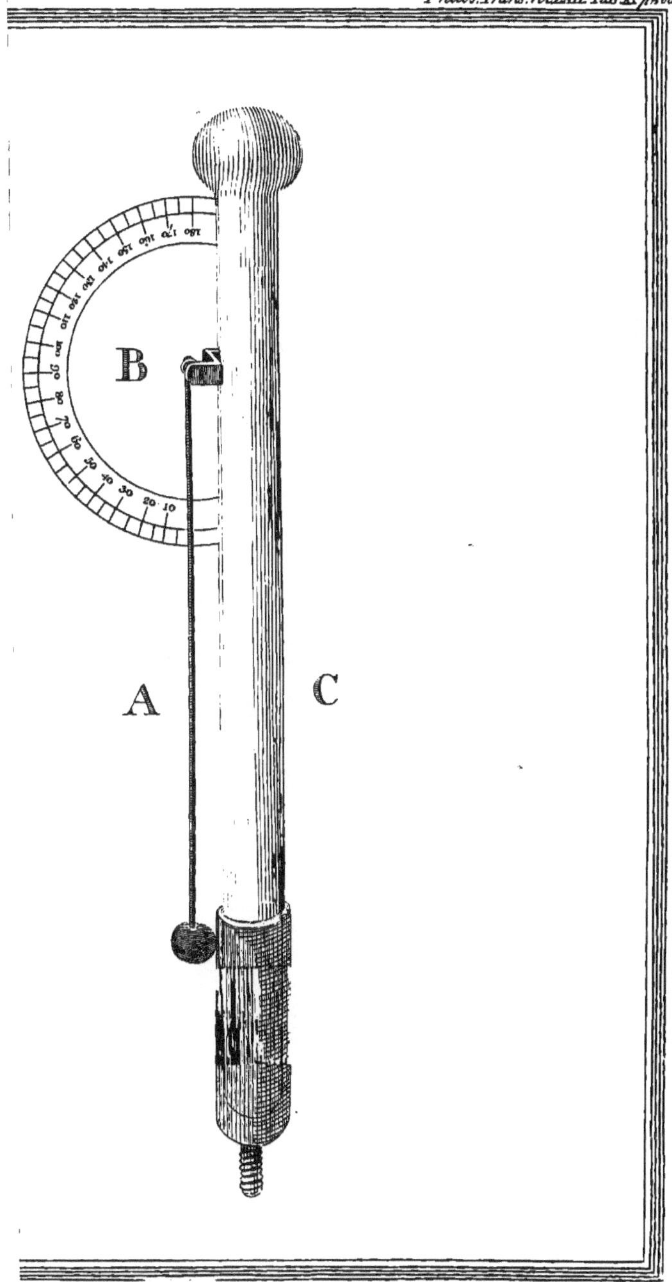

Basire sc.

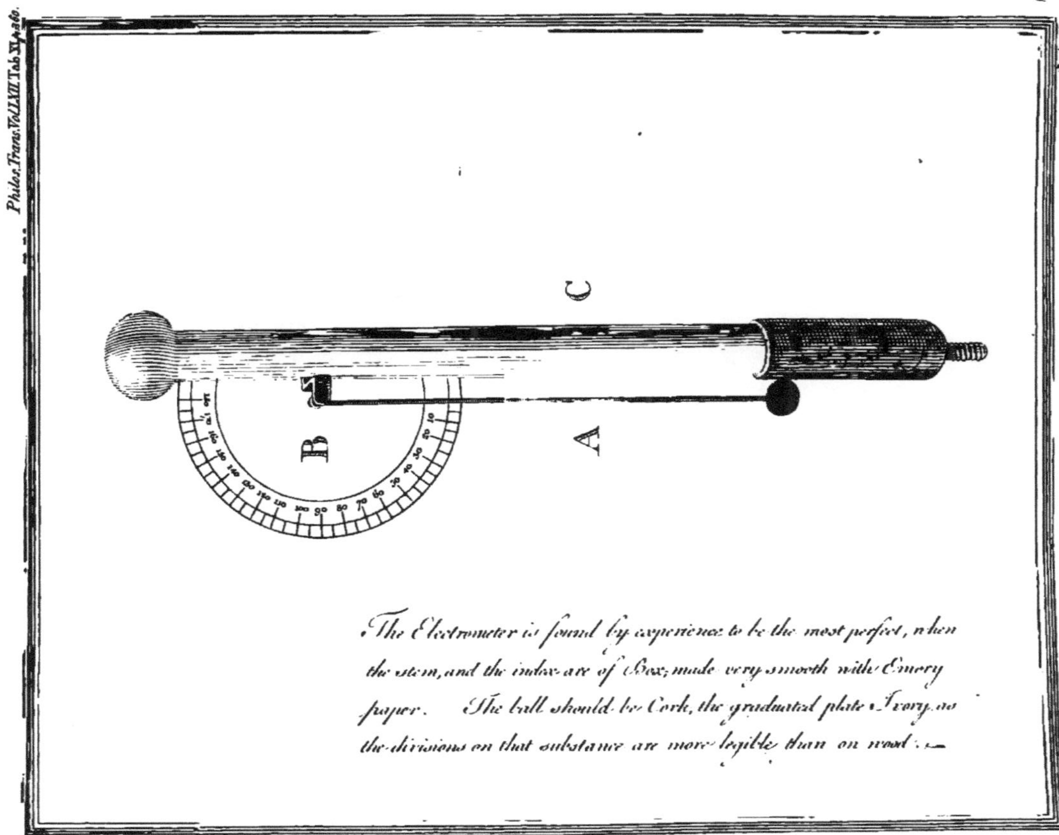
B
A
C
The Electrometer is found by experience to be the most perfect, when
the stem, and the index are of Box; made very smooth with Emery
paper. The ball should be Cork, the graduated plate Ivory, as
the divisions on that substance are more legible than on wood.
Basire sc.

exactneſs, to the degree in which the body to which it is connected is electrified; but this inconvenience is eaſily remedied, by moiſtening the ſtem and the index, for the ſemicircle cannot be too dry.

I find by experience, that this electrometer anſwers all the purpoſes I have mentioned, with the greateſt eaſe and exactneſs. I am now ſure of the force of any exploſion before a diſcharge of a jar or battery, which I had no better method of gueſſing at before, than by preſenting to them a pair of Mr. Canton's balls, and obſerving their divergency at a given diſtance; but the degree of divergency was ſtill to be gueſſed at by the eye, and the balls can only be appſied occaſionally; whereas this inſtrument, being conſtantly fixed to the prime conductor or the battery, ſhews, without any trouble, the whole progreſs of the charge; and, remaining in the ſame ſituation, the force of different exploſions may be aſcertained with the utmoſt exactneſs before the diſcharge.

If a jar be loaded with poſitive electricity, and I want to know the exact time when, by attempting to charge it negatively, it firſt becomes diſcharged, I ſee every ſtep of its approach to this ſtate by the falling of the index; and the moment I want to ſeize, is the time when it has got into a perpendicular ſituation, which may be obſerved, without the leaſt danger of a miſtake. Accordingly I find that, in this caſe, not the leaſt ſpark is left in the jar. If I continue the operation, the index, after having gained its perpendicular poſition, begins to advance again, and thereby ſhews me the exact quantity of the oppoſite electricity that it has acquired.

Conſidering the admirable ſimplicity, as well as the great uſefulneſs of this inſtrument, it is ſomething ſurprizing that the conſtruction ſhould not have occurred to ſome electrician before this time. Nollet's and Mr. Waits's invention of threads, projecting ſhadows upon a graduated board, reſembled this apparatus of Mr. Henly's, but was a poor and awkward contrivance in compariſon with it; nor was Richman's gnomon, though a nearer approach to this conſtruction, at all comparable to it; and the ingenious author of it had no knowledge of either of thoſe methods when he hit upon this.

I have made a receptacle for this inſtrument in my prime conductor, and I have alſo a pedeſtal in which I can fix it; and by means of which I can very conveniently place it on the wires of a battery.

In either of thoſe ſituations it anſwers almoſt every purpoſe of an electrometer, without removing it from its place.

I doubt not that you and all other electricians will join with me in returning our hearty thanks to Mr, Henly for this excellent and uſeful inſtrument.

Many of the effects of my battery, in breaking of glaſs, and tearing the ſurface of bodies, Mr. Henly performs by a ſingle jar, only increaſing the weight with which the bodies are preſſed, while the exploſion is made to paſs cloſe under them.

By this means he raiſes exceeding great * weights, and ſhatters ſtrong pieces of glaſs into thouſands of the ſmalleſt fragments; he even reduces thick plate glaſs by this means to an impalpable powder. But

* Frequently ſix pounds Troy.

what

what is moſt remarkable is, that when the pieces of glaſs are thick, and ſtrong enough to reſiſt the ſhock, they are marked by the exploſion, with the moſt lively and beautiful colours, generally covering the ſpace of about an inch in length, and half an inch in breadth.

In ſome of the pieces which he was ſo obliging as to ſend me, theſe colours lie all intermixed and confuſed; but in others I obſerve them to be diſpoſed in priſmatic order, in lines parallel to the courſe of the exploſion, and in ſome (as N° 1.) I have counted three or four diſtinct returns of the ſame colour.

He has lately informed me, that, ſince he ſent me this piece, he has ſtruck theſe priſmatic colours into another maſs of glaſs, in a ſtill more vivid and beautiful manner, the colours ſhooting into one another. This effect, he ſays, was produced by making a ſecond exploſion, without moving any of the apparatus after the firſt.

When the glaſs in which theſe colours are fixed is examined, it is evident that the ſurface is ſhattered into thin plates, and that theſe give the colours, the thickneſs of them varying regularly, as they recede from the path of the exploſion.

In the middle of theſe coloured ſpots (as in N° 2.) ſome of theſe thin plates, or ſcales, are ſtruck off, I ſuppoſe by the force of the exploſion; and with the edge of a knife they are all eaſily ſcraped away, when the ſurface of the glaſs is left without its poliſh (as in N° 3.)

The piece of glaſs on which I have marked theſe numbers, as well as that on which he has ſtruck the

 colours

colours in a ſtill more beautiful manner, Mr. Henly will preſent to the Royal Society, for the inſpection of the members.

Beſides theſe improvements, Mr. Henly has likewiſe, in a very ingenious manner, diverſified ſeveral of the more entertaining experiments in electricity, particularly in his imitation of the effects of earthquakes by the lateral force of exploſions; and he has alſo hit upon ſeveral curious facts, that, unknown to him, had been obſerved before by others: the following particular, however, I believe is new, exciting a ſtick of ſealing wax, and uſing a piece of tin foil for the rubber, he found that it would electrify poſitively, as well as glaſs rubbed with ſilk and amalgama.

Wiſhing we had more ſuch fellow labourers as Mr. Henly, I am,

DEAR SIR,

Your obliged

humble ſervant,

Leeds,
Oct. 26, 1770.

J. Prieſtley.

XXVII. *Me-*

Read May 28, 1772.

XXVII. *Meteorological Obſervations at* Ludgvan *in* Mount's-Bay, Cornwall, 1771: *By* William Borlaſe*, *D. D. F. R. S. Communicated by Dr.* Jeremiah Milles, *Dean of* Exeter, *and F. R. S.*

Month.	Barometer.	State of the Weather and Wind.	Fahrenheit's Thermom.	Ombr.
			Monthly Med. of heat for each day.	Inches.
January	Higheſt 23 30,5 Loweſt 19 29,0	The 1ſt at night a violent ſtorm, and rain till midnight. On the 2d at 8 P. M. a violent ſtorm, which continued all night; ſtormy the 3d and 4th; ditto the 26th at night; wind Weſterly. On the 10th at night, after hail ſhowers, a great fall of ſnow; the 11th great ſnow falling, with ſtormy blaſts; the 12th deep ſnow and more falling, with froſt; deep ſnow, and hard froſt, the 13th, 14th, 15th, 16th, 17th, 18th, and 19th, ſnow lying deep, but the froſt more gentle and the thaw came on; the 20th P. M. it thawed faſt; on the 21ſt in the afternoon, the froſt and ſnow was all gone; the reſt of the month moſtly miſts with ſome hard ſhowers of rain. Wind, during the ſtorms, miſts, and rain, Weſterly for 16 days; during the cold, Eaſt, and Eaſt North Eaſt.	Higheſt 1 50 Loweſt 17 $27\frac{1}{2}$ } $39\frac{5}{31}$	$3\frac{300}{1000}$

* This is the laſt paper of this kind, which the Society will receive from the excellent author of the Natural Hiſtory of Cornwall, and ſeveral other learned works; death having, though at an advanced age, put a period to a life divided between the purſuit of uſeful and experimental knowledge, and the faithful diſcharge of every moral and religious duty. M. M.

Month

Month.	Barometer.	State of the Weather and Wind.	Fahrenheit's Thermom.	Ombr.
			Med.	Inches.
February	Higheſt 3 30,16 Loweſt 25 28,87	Calm, the 3d, 4th, 5th, 6th, 7th, 9th, 14th, 15th, 17th, 18th, 21ſt, 22d; hard froſt with ſome ſnow on the 9th, 10th, 11th, 12th, 13th, 14th. It then thawed, and the reſt was hazy, miſty, ſhowery, with ſome high winds on the 15th and 27th. Wind, during the cold, Eaſt and North, the reſt South for 18 days.	Higheſt 21 52 Loweſt 11 30 } $43\frac{12}{28}$	1,500
March	Higheſt 19 30,6 Loweſt 12 29,15	Calm, the 3d, 4th, 5th, 6th, 7th, 8th, 15th, 18th, 19th, 20th, 21ſt, 27th, 31ſt. Froſt 6th, 7th, 23d, 25th, 28th; hail, ſnow, and ſleet 23d, 24th, 25th, 27th, 28th. Stormy the 1ſt, 13th. Wind 27 days from the Eaſt mixed equally with North and South.	Higheſt 13 $49\frac{1}{2}$ Loweſt 25 30 } $41\frac{27}{31}$	2,900
April	Higheſt 18 30,27 Loweſt 30 29,50	Calm, the 1ſt, 5th, 9th, 10th, 12th, 13th, 14th, 19th, 20th, 21ſt, 23d, 26th, 27th, 29th, 30th. Hailed, ſnow lying only 2 days, *viz.* the 15th and 16th. Reſt, moſtly fair, and dry. Wind 18 days from the Eaſt; the reſt mixed, and changeable.	Higheſt 22 53 Loweſt 16 35 } $46\frac{4}{30}$	0,900
May	Higheſt 23 30,8 Loweſt 7 29,18	Calm, 1ſt, 2d, 3d, 4th, 5th, 8th, 9th, 10th, 11th, 12th, 13th, 14th, 15th, 16th, 17th, 18th, 19th, 20th, 21ſt, 22d, 23d, 24th, 25th, in all 23 days. Stormy only on the 27th. Wind Southerly 23 days; the reſt not ſo fixed.	Higheſt 14 $65\frac{1}{2}$ Loweſt 1 45 } $53\frac{12}{31}\frac{1}{8}$	2,250

Month.

Month.	Barometer.	State of the Weather and Wind.	Fahrenheit's Thermom.	Ombr.
June	Higheſt 3 30,15 Loweſt 1 29,54	Calm, 3d, 4th, 5th, 6th, 7th, 8th, 9th, 10th, 11th, 12th, 13th, 14th, 15th, 20th, 21ſt, 22d, 23d, 24th, 25th, 26th, 27th, 28th, 29th, 30th (in all 24 days) the wind variable and mixed, but the weather remarkably ſettled, fair, and pleaſant. On the 28th however, there was moſt violent thunder, lightning, and a flood of rain, at the towns of Penryn and Falmouth, 20 miles diſtant from Mount's-Bay to the Eaſt; but in Mount's-Bay, the air was cloudy, and only ſome diſtant thunders; the lightning was ſcarce viſible, and not a drop of rain.	Med. Higheſt 27 $71\frac{1}{2}$ } $58\frac{9}{30}\frac{1}{4}$ Loweſt 3 49 }	Inches. 0,200
July	Higheſt 14 30,15 Loweſt 31 29,50	Calm, 1ſt, 2d, 3d, 4th, 5th, 6th, 7th, 10th, 13th, 14th, 15th, 16th, 17th, 18th, 23d, 24th, 25th, 27th, 28th, 29th, 30th, the reſt mixed. Wind 24 days from the Weſt, mixed moſtly with the South. N. B. As we had a moſt unuſual run of dry weather here in Cornwall; in other parts of the world, they had altogether as extraordinary a glut of rain. See, for particulars, the publick papers, from Berlin, Dreſden, Hamburgh, and Vienna, &c, in Europe, and from Virginia in North America, where their inundations have not been remembered ſo deſtructive.	Higheſt 17 72 } $61\frac{9}{31}\frac{1}{4}$ Loweſt 31 $54\frac{1}{2}$ }	0,720

Month.

Month.	Barometer.	State of the Weather and Wind.	Fahrenheit's Thermom.	Ombr.
			Med.	Inches.
August	Highest 29 30,0 Lowest 19 29,40	Calm, 1st, 2d, 3d, 4th, 5th, 6th, 15th, 17th, 22d, 23d, 26th, 27th, 28th, 29th, 30th, 31st ; the rest misty, fogs, and rains, with some gales interspersed. Wind 28 days from the West, mixed, nearly equal with North and South.	Highest 31 66 Lowest 1 54 } $58\frac{24}{31}$	3,633
September	Highest 28 30,6 Lowest 20 29,30	Calm, 2d, 3d, 5th, 6th, 10th, 11th, 12th, 13th, 14th, 15th, 16th, 17th, 18th, 24th, 25th, 26th, 28th ; the rest hazy, cloudy, windy, mixed with rain. Wind mostly West, mixed with the North.	Highest 1 61 Lowest 26 $46\frac{1}{2}$ } $55\frac{23}{30}$	3,400
October	Highest 31 30,33 Lowest 13 28,85	Calm, only 8 days A. M. only 2 days P. M. ; the rest rainy, windy, stormy. A violent storm on the 13th and 14th ; the extream on the 14th at 10 P. M. wind South West. N. B. on the 13th, at Caton near Lancaster, happened the greatest inundation in the memory of man. Wind 23 days from the West, mixed mostly with the South.	Highest 8 57 Lowest 31 43 } $51\frac{27\frac{1}{4}}{31}$	4,550
November	Highest 18 30,40 Lowest 11 29,40	Calm, 1st, 2d, 3d, 4th, 5th, 6th, 7th, 8th, 9th, 13th, 18th, 19th, 20th, 22d, 25th, 26th, 27th. On the 11th, 12th, 15th, 16th, stormy with rain and showers. Wind 24 days from the West, mixed mostly with the South. N. B. This month was very dry in Cornwall ; but by the incessant rains in the middle of it, from the 15th to the 17th, such inundations happened at Newcastle, Durham, Barnard-castle, and near Carlisle, by the breaking out of Solway-moss, as have never been known so destructive.	Highest 16 55 Lowest 10 39 } $47\frac{24\frac{3}{4}}{30}$	1,450

Month.	Barometer.	State of the Weather and Wind.	Fahrenheit's Thermom.	Ombr.
December	Highest 30 30,4 Loweſt 7 28,40	Calm, 1ſt, 2d, 8th, 13th, 17th, 18th. On Friday the 6th (new-moon at 8 A. M. wind South by Eaſt, and South Eaſt) about 8 P. M. it blew a violent ſtorm, and ſetting full into Mount's-Bay at ſpring-tide, the ſea was ſo high, and furious, that it demoliſhed houſes, cellars, boats, and walls, wherever it reached. The four towns on the ſhore all ſuffered; and it has been calculated, that not leſs than 5000 £. damage was done here that night, beſides ſhips loſt. It reached to the Eaſtward; and at Plymouth, about 60 miles off, they reckoned the tide was higher by ten feet than uſual. The remainder of this month was ſhowery, rainy, windy. Wind Weſterly 24 days, mixed moſtly with the South.	Med. Higheſt 11 54 } $46\frac{29}{31}\frac{1}{4}$ Loweſt 30 36 }	Inches. 5,350

The whole Rain fallen in this Year 1771, at Ludgvan, a very dry Year, } $30,\frac{153}{1000}$

XXVIII. Account

XXVIII. *Account of ſeveral Quadrupeds from* Hudſon's Bay *, *by Mr.* John Reinhold Forſter, *F. R. S.*

Read May 21, 1772.

1. Arctic Fox, Penn. Synopſ. of Quadr. p. 155. n. 113. *Canis Lagopus*, Linn.

Severn River.

A moſt beautiful ſpecimen in its ſnowy winter furr; this animal ſeems to be lower on its legs than the common fox, and is prodigiouſly well ſecured againſt the intenſe cold of the climate, by the thickneſs and length of its hairs, which are at the ſame time as ſoft as ſilk.

* Among the occaſional advantages, which the obſervations of the laſt Tranſit of Venus have procured, that of receiving uſeful informations from, and ſettling correſpondencies in, ſeveral parts of the world, is not the leaſt conſiderable. From the factory at Hudſon's Bay, the Royal Society were favoured with a large collection of uncommon quadrupeds, birds, fiſhes, &c. together with ſome account of their names, place of abode, manner of life, uſes, by Mr. Graham, a gentleman belonging to the ſettlement on Severn River; and the governors of the Hudſon's Bay Company have moſt obligingly ſent orders, that theſe communications ſhould be from time to time continued. The deſcriptions contained in the following papers were prepared and given by Mr. Forſter, before his departure on an expedition, which will probably open an ample field to the moſt important diſcoveries. M. M.

The

The account ſent along with it from Severn River ſays, that theſe white foxes are ſilly, inoffenſive animals; and are known to ſtand by, whilſt a trap is baited for them, into which they put their heads immediately: they will, when pinched by hunger, devour thoſe of their own kind, which are already caught in theſe traps. But the moſt curious circumſtance is, their migration to the Northward and the Eaſtern coaſts of the bay; for though a few of them are caught every year near York fort and Churchill river, yet, once in three or four years, they come in great numbers; and ſeveral hundred of their furrs are ſent to England in that plentiful ſeaſons, which always begins in November, and ends in April. The ſpecimen ſent is full grown, and its furr quite in ſeaſon.

2. Lesser Otter. Penn. Syn. Quadr. p. 239. n. 174. *Muſtela Lutreola* Linn. Syſt. Nat. 66. Faun. Suec. N° 13.

Severn River.

I am ſtill dubious, whether this animal ought to be looked upon as the ſame with the leſſer otter of Europe and Aſia; many circumſtances ſeem to prove this identity; but ſome, ſuch as the want of webs, which I could not diſcover between the toes, and the white ſpot on the neck, will not admit of it. I have, therefore, ſubjoined a deſcription of this creature at the end of this article. The natives of Hudſon's Bay call this quadruped

 Jackaſh;

Jackaſh; Mr. Graham from Severn river ſays, that it harbours about creeks, and lives on fiſh, like the otter; it travels very ſlowly, and has from four to ſeven young at a time; in ſize it equals the marten; its length is about 16 inches; its whole body is covered with ſhining dark brown hairs, which lie very cloſe, and ſeem perfectly convenient for an amphibious animal; under theſe brown hairs the woolly hairs are tawny, the whole under-jaw is encompaſſed by a ſtripe of white hairs, and a little irregular ſpot of the ſame colour appears in the middle of the throat; the feet are quite covered with hair to the very nails, which are ſmall, five on each foot, and of a whitiſh ſemipellucid colour; the tail is pretty well beſet with hair, though not buſhy, and much blacker than the reſt of the body; it is about half as long as the whole animal.

3. PINE MARTEN. Penn. Syn. Quad. p. 216. n. 155. *Muſtela Martes (Abietum)*. Linn.
Severn River. Male and Female.

Theſe ſeem to be a variety of the yellow-breaſted marten, Br. Zool. I. 81. their colour, eſpecially in the females, being much paler than that deſcribed in Mr. Pennant's works. The male is of a cheſnut brown, the female a bright tawny yellow; the former has here ſome dark brown hairs, the latter in the ſame manner has ſome bright bay hairs. They both have white cheeks, and white tips of the ears. Their furrs are very full of hair, proper to preserve them from the cold. The tail in both sexes is bushy, and darker than the rest of the body; in the female indeed it is tawny, with a black tip; in both it is shorter than described by Mr. Pennant, Mr. Brisson, and others, and was perhaps mutilated. This species feeds on mice, rabbits, &c. though it will not touch a dead mouse which is put as a bait in a trap, and therefore the inhabitants are obliged to make use of a partridge's head, or the like, for that purpose. If pursued with noise, it immediately gets up into a tree. Some gentlemen have unsuccessfully attempted to tame these creatures, and those kept in cages with that view have been observed to be troubled with epileptick fits. Numbers of them are caught at Hudson's Bay in traps made of small sticks. They burrow under ground, and bring forth from four to seven young at a time.

4. STOAT AND ERMINE. Penn. Syn. Quad. p. 212. n. 151. α. β. *Mustela Erminea.* Linn.

Severn River, Albany Fort.

One in the summer and another in the winter dress. The natives about Albany call them *Sic-cuse-sue*, but it is not known why they give them that name. They feed on mice, small birds, all sort of fish, flesh, and fowl.

5. COMMON WEESEL. Penn. Syn. Quadr. p. 211. n. 150. *Mustela nivalis.* Linn.

One in its winter dress, length 7 inches, tail about 1 inch, perhaps mutilated; it is quite white, but the

the coat is mixed here and there with a brownifh hair, efpecially in the tail. Another in the fummer coat, the fame as our weefel.

6. SKUNK. Penn. Syn. Quadr. p. 233. n. 167. Kalm's Travels, I. 273. tab. I.

It anfwers to Mr. Pennant's defcription, except that the white ftripe on the head is not connected with that on the back, and that the brown area, which is left between the two white ftripes on the back, is broader than he defcribes it.

7. CANADA PORCUPINE. Penn. Syn. Quadr. p. 266. n. 196. *Hyftrix dorfata.* Linn.

Severn River.

It agrees perfectly with the defcriptions. Thefe animals live among the pine trees, of which the bark is their food in winter, as willow tops and the like are in fummer. They copulate in September, and bring forth only one young the firft week in April. During winter they feldom travel above five hundred yards, fo that one is always fure of finding a porcupine, as foon as one meets with a tree that has been frefh ftripped of its bark. The longeft quills of an old porcupine are about five inches long. The Europeans are very fond of the flefh of thefe animals, as it taftes, when roafted, exactly like that of a fucking pig. Their bones in winter have a greenifh yellow colour, perhaps owing to their continually feeding on the bark of pine trees. It is known that

that the bones of animals will become red by their feeding on madder.

8. BEAVER. Penn. Syn. Quadr. p. 255. n. 190. *Castor Fiber.* Linn.

Churchill River, N° 1.

A most beautiful specimen, in high preservation, and in full season; the furr is of a fine jetty black: the skull of another has likewise been sent. There is a great similarity in the conformation of the cutting teeth of this and the preceding quadruped (the porcupine); only the latter has them longer.

9. MUSK-BEAVER. Penn. Syn. Quadr. p. 259. n. 121. *Castor Zibethicus.* Linn.

Musquash. Severn River.

It frequents the plains, builds a house like the beaver, brings forth from five to seven young at a time, and feeds on poplars, willows, and grass.

10. ALPINE HARE. Penn. Syn. Quadr. p. 249. n. 185. *Lepus timidus.* Linn. Kalm's Trav. into N. Amer. III. p. 59.

York Fort.

A fine specimen, in its compleat winter furr, being quite white, except the ears, which have black tips. It is much larger than the following animal. The common hare, *Penn. Syn. Quadr.* does not seem to be a native of America.

11. AME-

11. AMERICAN HARE, called Rabbit at Hudſon's Bay. Kalm's Trav. into N. Amer. I. 105. II. 45. Severn and Churchill Rivers.

This ſpecies, which has been improperly called Rabbit, perhaps becauſe it is leſs than the hare, is certainly new, and was never deſcribed before, except by Kalm in his travels through North America, Vol. I. 105. II. 45. The account he there gives correſponds with that of Mr. Graham, and with the ſpecimen now in the Royal Society's collection. Theſe animals are numerous at Hudſon's Bay; they do not burrow under ground, but live ſummer and winter under windfalls and roots of trees. They do not migrate, but always keep about the ſame place, unleſs diſturbed. They breed once or twice a year, and have five to ſeven young at a time: their weight is from 3 to 4½ pounds. Their fleſh is not ſo white and delicate as that of the common rabbit, but yet is good food in ſummer and winter. Great numbers of them are annually caught in the following manner: as they always are uſed to go one particular path, the Engliſh and natives lay young trees acroſs it, forming a hedge, in which there is an opening for the creature to go through; in this place they fix a ſnare, made of braſs wire, packthread, or the like, faſtened with a ſlipping knot to a croſs piece, the end being tied to an elaſtic pole; ſo that when the animal puts its head into

into the ſnare, the knot is drawn from the croſs piece above, and the pole flying up, immediately ſuſpends the animal in the air.

The proper characteriſticks of this ſpecies ſeem to be,

1. Its ſize, which is ſomewhat bigger than a rabbit's, but leſs then that of the Alpine or leſſer hare.
2. The proportion of its limbs, its hind feet being longer in proportion to the body than thoſe of the rabbit and the common hare. Vide the Hon. Daines Barrington's, V.P.R.S. letter to Dr. Watſon on this new ſpecies of hare, in this volume, p. 6.
3. The tips of the ears and tail, which are conſtantly grey not black. Kalm's Trav. II. p 45.

Perhaps ſome other characters might be aſcertained, if the animal was brought over in its perfect ſummer furr; for all the ſpecimens in the Royal Society's Muſeum are either entirely in their winter dreſs, or in a changing condition. Mr. Kalm mentions, that thoſe which are found in New Jerſey, where the climate is much more mild than at Hudſon's Bay, keep the ſame grey colour both ſummer and winter; that in ſpring they breed in hollow trees, but in ſummer in the graſs; that, when purſued, they immediately take refuge in hollow trees, whence they are driven out by crooked ſticks, ſmoak, &c.; laſtly, that they do much miſchief to cabbage fields and orchards, by eating the cabbage plants, and

 the

the bark of the apple trees, feeding only by night, as the common hare.

12. QUEBEC MARMOT; Penn. Syn. Quadr. p. 270. n. 199.

Churchill River, N° 5.

This creature is called a ground ſquirrel, at Churchill fort; it differs much in ſize from that deſcribed in the Syn. Quadr. being much leſs than a rabbit, perhaps it is a young one. I took down the following deſcription, as I did not find it exactly correſponding with that of the Canada marmot. The noſe is blunt, the ears are ſhort and roundiſh, the top of the head cheſnut, back all over ſprinkled with whitiſh, black, and yellowiſh brown: the legs and whole underſide of the animal are of a bright ferruginous colour; the tail is very ſhort, and black at the tip. The length of the animal from the noſe to the beginning of the tail is about 11 inches, that of the tail 3 inches. Its toes on the fore feet 4, hind feet 5.

13. COMMON SQUIRREL. Penn. Syn. Quadr. p. 279. n. 206. *Sciurus vulgaris*, Linn.

A variety of the common ſpecies, being ſomewhat inferior in ſize, having a ferruginous back and grey belly, a ſhorter tail than the common European ſort, of a fine ferruginous red, edged only with black. This animal lives in pine trees, of which the cones are its food; it lies dormant the greater part of the winter.

14. GREATER FLYING SQUIRREL.
Severn River.

It is equal in ſize, if not bigger than the common ſquirrel; has pretty long hairs, duſky at bottom, tawny brown at the very tips only; and diſpoſed ſo that the back appears wholly of that reddiſh brown colour; the tail is very buſhy, ſomewhat compreſſed, but not pinnated (i. e. with the hairs diſpoſed horizontally on each ſide of it, as for example in the common ſquirrel), it is browniſh on the upperſide with a duſky tip, of a yellowiſh white below; the whole underſide of the animal has the ſame yellowiſh white colour. The membrane reaches from the forefeet to the hindfeet, without extending to the ears: it is found in James's Bay, about 51° north latitude.

This is perhaps Linneus's *Sciurus volans*, and the ſame with the flying ſquirrel of the Arctick parts of Europe. Mr. Briſſon ſeems to have confounded this, and the little Virginian ſquirrel together, and his quotations are quite confuſed. Linneus's *Mus volans* certainly is a variety of the little flying ſquirrel, of the milder parts of North America, New York, Pennſylvania, Virginia, which is vaſtly different from this in ſize and colour.

15. A SMALL ANIMAL, called a Field Mouſe.
Churchill River.

A ſpecimen in very bad preſervation, wanting legs, tail, &c. which makes it impoſſible to de-

 termine

termine of what ſpecies it is; its ſize is ſomewhat ſuperior to that of a mouſe, its colour duſky, mixed wirh tawny brown, and dirty white on the belly; its head is broad, like that of the ſhort-tailed field mouſe, and has a duſky line in the middle between the eyes, which extends, though rather indiſtinctly, all along the back; its ears are very ſmall and roundiſh.

16.

This is likewiſe a very bad mutilated ſpecimen, leſs than the common mouſe, duſky and brown above, and whitiſh below; its ears are pretty large and prominent.

17. FIELD MOUSE. Penn. Syn. Quadr. p. 302. n. 230. *Mus Sylvaticus*, Linn.

Two ſpecimens; the deſcriptions anſwer pretty well, the ears are large and round, the tail is very long and whitiſh below.

18. SHORT-TAILED MOUSE. Penn. Syn. Quadr. p. 305. n. 233. *Mus terreſtris*, Linn. Le Campagnol de Buffon.

Mr. Pennant's admeaſurements do not quite anſwer, but M. d'Aubenton's coincide.

19. FOETID SHREW. Penn. Syn. Quadr. p. 307. n. 235. *Sorex Araneus*, Linn.

The ſpecimen is much blacker on the back than the European Shrew, its ſides are reddiſh brown.

20. SHREW; two ſpecimens.

The colour is of a duſky grey above, and a dirty white or yellowiſh below; the noſe is very long and ſlender; the length from the noſe to the tail, in the one ſpecimen is 2¼, in the other almoſt 2 inches; the tail is about an inch and half long, thinly beſet with hairs, brown above, and yellowiſh below. If this ſpecies had no tail, I ſhould take it to be the minute Shrew, which the Rev. Mr. Laxman found in Siberia, and which is the *Sorex minutus*. Linn.

XXIX. *An Account of the Birds ſent from* Hudſon's Bay; *with Obſervations relative to their Natural Hiſtory*; *and* Latin *Deſcriptions of ſome of the moſt uncommon. By* J. R. Forſter, *F. R. S.*

Read June 18—25, 1772.

I. Land-Birds.

1. { Accipitres
Rapacious. Faun. Am. Sept.

1. Falco, } 1. Columbarius. 128. 21. Pigeon Hawk.
Falcon. } Faun. Am. Sept. p. 9. Cateſby I. t. 3.
Epervier de la Caroline. Briſſon I. p. 378.
Severn river, N° 19.

This ſpecies is called a *ſmall-bird hawk* at Hudſon's Bay. It is migratory, arriving near Severn River in May, breeding on the coaſt, and then retiring to a warmer climate in autumn. It feeds on ſmall birds; and, on the approach of any perſon, will fly in circles, making a hideous ſhrieking noiſe. The breaſt and

and belly are yellowiſh, with brown ſtreaks, which are not mentioned by the ornithologiſts, though their deſcriptions anſwer in other reſpects. It weighs ſix ounces and a half, its length is 10½, the breadth 22½. Cateſby's figure is a very indifferent one.

FALCO, 2. Spadiceus. *New Species.* Chocolate Falcon. Faun. Am. Sept. p. 9.

This ſpecies, at firſt ſight, bears ſome reſemblance to the European Moor Buzzard, or *Aeruginoſus*, Linn. but is much leſs, and wants the light ſpots on the head and ſhoulders. No number or deſcription was ſent along with it.

FALCO, 3. Sacer, Briſſon, I. p. 337. Sacre de Buffon, Oiſeaux, (edition in 12mo.) Tom. II. p. 349. t. 14. Faun. Am. Sept. p. 9.

Severn River, N° 16.

Speckled Partridge Hawk, at Hudſon's Bay. The name is derived from its feeding on the birds of the Grous tribe, commonly called partridges, at Hudſon's Bay. Its irides are yellow, and the legs blue. It comes neareſt the *Sacre* of Briſſon, Buffon, and Belon; but Buffon ſays it has black eyes, which is very indiſtinct; for the irides are black in none of the falcons, and in few other birds; and the pupil, if he means that, is black in all birds. It is ſaid, by Belon, to come from Tartary and Ruſſia, and is, therefore, probably a northern bird. It is very voracious and

and bold, catching partridges out of a covey, which the Europeans are driving into their nests. It breeds in April and May. Its young are ready to fly in the middle of June. Its nests, as those of all other falcons, are built in unfrequented places; therefore, the author of the account from Severn River could not ascertain how many eggs it lays; however, the Indians told him it commonly lay two. It never migrates, and weighs $2\frac{1}{2}$ pounds; its length is 22 inches, its breadth 3 feet.

2. STRIX, Owl. } 4. Brachyotos. The short-eared Owl. Brit. Zoology, folio, plate B. 3. octavo, I. p. 156. Faun. Am. Sept. 9.
Severn River, N° 17 and 64.

Mouse Hawk at Hudson's Bay. It answers the description and figure in the British Zoology; but its ears or long feathers do not appear. The smallness of the head has, probably, given occasion to call it a hawk, though it does not fly about in quest of prey, like other hawks (as the account from Severn River says); it sits quiet on the stumps of trees, waiting mice with all the attention of a domestic cat, being an inveterate enemy of those little animals. It migrates southward in autumn; and breeds along the coast. Its irides are yellow. Its weight is 14 ounces; its length 16 inches, the breadth 3 feet.

STRIX, 5. Nyctea. 132. 6. Snowy Owl. Faun. Am. Sept. 9.

Churchill River, N° 7. White Owl.

It ſeems to be in its winter dreſs, as it is intirely white. The feet are covered with long white hair-like feathers to the very nails, but there are none on the ſoles or under parts of the toes.

STRIX, 6. Funerea. 133. 11. Canada Owl. Faun. Am. Sept. 9.

Severn River, N° 13. Churchill River, N° 11.

Cabeticuch, or *Cabaducutch*, is the Indian name of this bird. Linneus's deſcription anſwers perfectly. The male, which in the claſs of birds of prey is generally ſmaller, is, however, in this ſpecies, larger than the female, according to the account from Severn River. Its colour is likewiſe much blacker, and the ſpots more diſtinct. The eyes are large and prominent; the irides of a bright yellow. The weight is 12 ounces; its length 17 inches, the breadth 2 feet. It has only two young at one hatching.

STRIX, 7. Paſſerina. 133. 12. Little Owl. Brit. Zool. Faun. Am. Sept. 9.

(The number belonging to this bird is loſt, but it is moſt probably that from Severn River, N° 15. called *Shipcmoſpiſh* by the natives).

The crown of the head is ſpeckled with white, as in the *Strix funerea*.

STRIX, 8. Nebulosa. *New species.* The grey Owl.
Severn River, N° 36.

This fine non-descript owl lives upon hares, ptarmigans, mice, &c. It has two young at a time. The specimen sent over is said to be one of the largest. It is not described by any author. Its weight is 3 pounds, length 16 inches, breadth 4 feet.

3. LANIUS, Shrike. } 9. Excubitor. 135. 11. Great Butcher-bird. Brit. Zool. Cinereous Shrike. Faun. Am. Sept.
Severn River, N° 11.

White Whiskijohn at Hudson's Bay. The specimen is a male; it weighs two ounces and a half, is seldom found on the coast, but frequent about a hundred miles inland; and feeds on small birds. It corresponds with ours in every respect.

II. { Picæ. Pies. Faun. Am. Sept.

4. CORVUS, Crow. } 10. Canadensis. 158. 16. Cinereous Crow. Faun. Am. Sept. 9.
Severn River, N° 9 and 10.

These birds are called *Whiskijohn* and *Whiskijack* at the Hudson's Bay. They weigh 2 ounces; and are 9 inches long, and 11 broad. Their eyes are black, and their feet of the same colour. Their characters correspond with the Linnean description. They breed early in spring; their nests are made of sticks and grass,

graſs, and built in pine trees; they have two, rarely three, young ones at a time; their eggs are blue; they fly in pairs; the male and female are perfectly alike; they feed on black moſs, worms, and even fleſh. When near habitations or tents, they are apt to pilfer every thing they can come at, even ſalt meat; they are bold, and come into the tents to eat victuals out of the diſhes. They watch perſons baiting the traps for martins, and devour the bait as ſoon as they turn their backs. Theſe birds lay up ſtores for the winter, and are ſeldom ſeen in January, unleſs near habitations; they are a kind of mock-bird; when caught, they pine away and die, though their appetite never fails them.

Corvus, 11. Pica. 157. 13. Magpie. Brit. Zool. Faun. Am. Sept. 9.

Albany Fort, N° 5.

It is called *Oue-ta-kee-aſke*, i. e. *Heart-bird*, by the Indians. It is a bird of paſſage, and rarely ſeen; it agrees, in all reſpects, with the European magpie, upon compariſon.

5. Picus, Woodpecker. } 12. Auratus. 174. 9. Gold-wing Woodpecker. Faun. Am. Sept. 10. Cateſby, I. 18.

Albany Fort, N° 4. the large Woodpecker.

The natives of America call this bird *Ou-thee-quan-nor-now*, from the yellow colour of the ſhafts of the quill and underſide of the tail feathers. It is a bird of paſſage; viſits the neighourhood of Albany Fort in April, leaves it in September; lays from four to fix eggs in hollow trees, feeds on fmall worms and other infects. Its defcriptions anfwer exactly.

Picus, 13. Villofus, 175. 16. Hairy Woodpecker. Faun. Am. Sept. 10. Catefby I. 19.

Severn River, N° 56.

The fpecimen fent over is a female, by its wanting the red on the head. The defcriptions of Linneus and Briffon agree; only the two middlemoft feathers are black, the next are of the fame colour, but have a white rhomboidal fpot near the tip; the next are black, with the upper half obliquely white, the very tip being black; the next after that are white, with a round black fpot on the inner fide clofe to the bafe, and the lower part of the fhaft is black, the outermoft feathers are quite white, the fhaft only at the bafe being black.

14. Tridactylus. 177. 21. Three-toid Woodpecker. Faun. Am. Sept.

Severn River, N° 8.

A female, weight 2 ounces, length 8 inches, breadth 13; eyes dark blue, legs black. It builds its neft in trees, lives in woods upon worms picked out of trees, is not very common at Severn River. The defcriptions anfwer.

III. Gallinæ,

III. { Gallinæ.
Gallinaceous. } Faun. Am. Sept.

6. Tetrao. { 15 Canadensis, 274. 3. } Faun. Am. Sept. 10.
Grous. { Canace, 275. 7. } Spotted Grous.
Gelinotte du Canada, male et femelle, Pl. enl. 131 et 132. Buffon Oiseaux II. p. 279. 4to. Brisson I. p. 203. t. 20. f. 1, 2, and p. 201. app. 10. Edwards, t. 118 and 71.

Severn River, N° 5. Woodpartridge.

These birds are all the year long at Hudson's Bay, and never change the colour of their plumage. The accounts from Hudson's Bay say, there is no material difference between the male and female; which must be a mistake, as they are really very different. Linneus's descriptions of the Tetrao Canadensis, and Canace, both answer to the specimens sent over, so that, after comparing them, I find they are only one and the same species. I suppose the dividing them into two, was occasioned by Brisson's and Edwards's descriptions, being taken from specimens sent from different parts of the continent of America, and perhaps caught at different seasons. Mr. de Buffon has, I find, the same opinion with me, and by comparing the drawings of Edwards, with those of the Planches enluminées, it is put beyond a doubt. These birds are very stupid, may be knocked down with a stick, and are frequently caught by the natives with a ſtick and a loop. In ſummer they are good eating; but in winter they taſte ſtrongly of the pine ſpruce, upon which they feed during that ſeaſon, eating berries in ſummer. They live in pine woods, their neſts are on the ground; they generally lay but five eggs.

Tetrao, 16. Lagopus, 274. 4. White Grous. Faun. Am. Sept. 10. Ptarmigan. Br. Zool. Lagopéde de la Baye de Hudſon. Buffon Oiſeaux II. p. 276. Edw. t. 72.

Severn River. N° 1—4. Willow-partridges.

The Hudſon's Bay ptarmigan has been ſeparated from the European in the Britiſh Zoology, and afterwards by M. de Buffon: however, I muſt own, I cannot yet find the differences which they aſſign to theſe ſpecies. They contend that the Hudſon's Bay bird figured by Edwards is twice as big as the European ptarmigan; Mr. Edwards, I think, does not intimate this, when he ſays, the bird is of a middle ſize, between partridge and pheaſant; he on the contrary ſuppoſes them to be the ſame ſpecies. The Britiſh Zoology, after Willoughby, ſays, the ptarmigan's length is 13¾ inches. The account from Severn River ſays it is 16¼ inches. The breadth in the Britiſh Zoology is ſaid to be 23 inches. The breadth in the Hudſon's Bay birds, according to the accounts from Severn River, is 23 inches. Willoughby's ptarmigan weighed 14 ounces; that in the Britiſh

Zool.

Zool. illuftr. t. 13. 19 ounces; that from the Hudfon's Bay (1½ ℔) 24 ounces. Thefe differences are of little confequence, and far from increafing the Hudfon's Bay bird to double the fize of the European. The Britifh Zoology fays, there is a difference in the fummer colours; but Mr. Edwards informs us, that he compared the Hudfon's Bay bird with the defcriptions of former ornithologifts, and found them to anfwer; he likewife affures us he had the fame bird from Norway. Therefore I cannot help diffenting from the Britifh Zoology, in this one particular, and thinking with Linneus and Briffon, that the European and Hudfon's Bay ptarmigans are the fame, efpecially as the colours vary very much in the different fexes and at different feafons. To this we may add the teftimony of a gentleman well verfed in natural hiftory, who, having had opportunities of comparing numbers of Hudfon's Bay and European ptarmigans, affured me that he did not fee any difference between them. They go together in great flocks in the beginning of October, living among the willows, of which they eat the tops (whence they have got the name of willow partridges): about that time they lofe their beautiful fummer plumage, and exchange it it for a fnowy white drefs, moft providently adapted by its thicknefs to fcreen them againft the feverity of the feafon, and by its colour againft their enemies the

the hawks and owls, againſt whoſe attacks they would otherwiſe find no ſhelter. Each feather is double, that is, a ſhort one under a long one, to keep them warm. In the latter end of March, they begin again to change their plumage, and have got their full ſummer dreſs by the end of June. They breed every where along the coaſt, and have from nine to eleven young at a time; making their neſts on the ground, generally on dry ridges. They are excellent eating, and ſo plentiful that ten thouſand have been taken at Severn, York, and Churchill Forts. The method of netting or catching them, is as follows: a net made of jack-twine, twenty feet ſquare, is laced to four long poles, and ſupported in front with the ſticks, in a perpendicular ſituation; a long line is faſtened to theſe ſupports, one end of it reaching to a place where a perſon lies concealed; ſeveral men drive the ptarmigans (which are as tame as chickens, eſpecially on a mild, ſnowy day), towards the net, which they run to, as ſoon as they ſee it. The perſon concealed draws the line, by which means the net falls down, and catches 50 or 70 ptarmigans at once. They are ſometimes rather wild, but grow better humoured (as Mr. Graham ſays) by being driven about, for they ſeldom forſake thoſe willows which they have once frequented.

TETRAO.

TETRAO. 17. Togatus, 275. 8. Shoulder-knot Grous. Groſſe Gelinotte du Canada. Pl. enl. 104. Briſſ. I. 207. t. 21. f. 1. Buffon Oiſeaux II. p. 287.

Severn River, N° 60 and 61. Albany Fort 1 and 2.

This bird anſwers the deſcriptions given of it by the ornithologiſts in all reſpects, and perfectly reſembles the figure in Briſſon, and in the Planches enluminées. It differs from Edwards's ruffed heathcock, t. 248. or Linneus's Tetrao umbellus, as the latter has not the ſhining black axillar feathers, or ſhoulder-knot, but a ferruginous one, is much leſs, and has brighter colours. M. de Buffon, however, thinks they are the ſame, and ſuſpects at the ſame time, that the bird which he calls la groſſe Gelinotte du Canada (and which is the ſame with the Society's ſpecimens) is the female of Mr. Edwards's bird, t. 248. This conjecture is deſtroyed by the ſpecimens now ſent from Hudſon's Bay, which by the accounts from thence are expreſſly ſaid to be males. The ſhoulder-knot grouſes bear the Indian name of *Puſkee*, or *Puſpuſkee*, at Hudſon's Bay, on account of the leanneſs and dryneſs of their fleſh, which is extremely white, and of a very cloſe texture, but when well prepared is excellent eating. They are pretty common at Mooſe Fort and Henly Houſe, but are ſeldom ſeen at Albany Fort, or to the northward of the above places. In winter they feed upon juniper tops, in ſummer on gooſe-berries, raſpberries, currants, cranberries, &c. They are not migratory, ſtaying all the year at Mooſe Fort; they build their neſts on dry ground, hatch nine young at a time, to which the mother clucks, as our common hen does; and on the leaſt appearance of danger, or in order to enjoy a comfortable degree of warmth, the young ones retire under the wings of their parent.

N. B. A ſpecimen, which is ſuppoſed to be either a young bird or a female, wants the blueiſh black ſhoulder-knot; but it is the ſame in all other reſpects.

TETRAO, 18. Phaſianellus. Linn. Syſt. Nat. Ed. X. p. 160. n. 5. Edw. 117. Longtailed Grous. Faun. Am. Septentr. 10.

Severn River, N° 6 and 7. Albany Fort, N° 3.

This bird, which Mr. Edwards has drawn plate 117, was by Linneus in the tenth edition of his Syſtem, ranged as a new ſpecies of grous or tetrao, by the ſpecific name of Phaſianellus (alluding to the name of Pheaſant which it bears at Hudſon's Bay, and likewiſe to its pointed tail). He afterwards in the new or twelfth edition of the Syſtem, p. 273. makes it a variety of the great Cock of the Wood, or Tetrao Urogallus, probably from the account in Mr. Edwards, that the male ſtruts very upright, is in general of a darker colour than the female, and has a gloſſy neck: Theſe circumſtances, however,e are not ſufficient to

bring

bring them under the ſame ſpecies, for it is known that the males of all the grous tribe, and indeed of moſt of the gallinaceous birds, are uſed to ſtrut in a very ſtately manner, and that the colours of their plumage are much more diſtinct than thoſe of the females. But the ſpecific difference alone, which Linneus aſſigns to the cock of the wood, abſolutely excludes our Hudſon's Bay ſpecies; he calls it Tetrao pedibus hirſutis, cauda rotundata, axillis albis. Whoever examines Mr. Edwards's figure, and the ſpecimens now in the Society's poſſeſſion, will find the tail very ſhort, but pointed, the two middle feathers being half an inch longer than the reſt, (Mr. Edwards ſays two inches) and the axillæ, or ſhoulders, by no means white: beſides this difference, the colour and ſize of the Hudſon's Bay bird are likewiſe vaſtly different from thoſe of the cock of the wood. Its length is 17 inches, its breadth 24, and, as Mr. Edwards juſtly ſays, it is ſomewhat bigger than the common pheaſant. The great cock of the wood is as big as a turky; and its female, which is much leſs, however far exceeds our bird, it being 26 inches long, and 40 broad. See Britiſh Zool. octavo, p. 200. The figures given of the female of the T. Urogallus, or great cock of the wood, in the Br. Zool. folio, plate M*, and the Planche enluminée 75, will ſerve upon compariſon as a convincing proof of the vaſt difference there is between the Hudſon's Bay pheaſant grous and the European cock of the wood. The figure, which Mr. Edwards has given of the former bird, does not exactly correspond with the Society's specimen, as he has represented the marks on the breast half-moon shaped, though they are heart-shaped as those on the belly in the dried bird; that is, they are white spots, with a pale brownish yellow cordated brim. Nor can I agree with Mr. Edwards, when he calls this bird the long-tailed grous from Hudson's Bay; for its tail is really very short, in comparison with that of other grouse, and its smallness and acuteness afford one of the most distinguishing characters of the species. The native Indians call these pheasant grouses, *Oc-kiss-cow:* they are found all the year long, amongst the small juniper bushes, of which the buds are their principal food, as also the buds of birch in winter, and all sorts of berries in summer. They never vary their colours; nor is there any great difference between the male and female, except in the caruncula or comb over the eye, which in the male is an inch long, and $\frac{1}{4}$ of an inch high. The account from Albany Fort adds, that the colour of the male is somewhat browner, and almost a chocolate on the breast. Their flesh is of a light brown, exceeding juicy, and they are very plump. They lay from 9 to 13 eggs; their young can run almost as soon as they are hatched; they make a piping noise somewhat like a chicken. The cock has a shrill crowing note, not very loud;

 but

but when diſturbed, or whilſt flying, he makes a repeated noiſe of cuck, cock. They are moſt common in winter at Albany Fort.

Before I leave the genus of grouſes, I muſt obſerve that their feet have a peculiarity, taken notice of by few authors; the toes, in ſeveral ſpecies, have on each ſide a row of ſhort flexible teeth, like thoſe of a comb; ſo that the toes appear pectinated. The ſpecies, which are known to have ſuch pectinated toes, are,

1. The great Cock of the Wood, *Tetrao Urogallus*, Linn.
2. The Black Cock, *T. Tetrix*, Linn.
3. The Spotted Grous, and { *T. Canadenſis*, *T. Canace*, Linn.
4. The Ruffed Grous, *T. Umbellus*, Linn.
5. The Shoulder-knot Grous, *T. Togatus*, Linn.
6 The Pheaſant Grous, *T. Phaſianellus*.
7. The Hazel Hen, *T. Bonaſia*, Linn.
8. The Pyrenæan Grous, *T. Alchata*, Linn.

This is a circumſtance, which ought to be attended to in all other ſpecies of grouſes, as it may in time afford a diſtinguiſhing character for a diviſion in this great genus; the ptarmigan, or *T. Lagopus*, Linn. is without theſe teeth.

IV. Co-

IV. {Columbæ.
Columbine. Faun. Am. Sept.

7. COLUMBA, Pigeon. } 19. Migratoria. 285. 36. Migratory Pigeon. Cateſb. I. 23. Kalm. II. p. 82. t. Paſſenger Pigeon, Faun. Am. Sept. 11.
Severn River, N° 63. Wood-pigeon.

Theſe pigeons are very ſcarce ſo far northward as Severn river, but abound near Mooſe-fort, and further inland to the ſouthward. Their common food are berries and juniper buds in winter; they fly about in great flocks, and are reckoned good eating. This account is confirmed by Kalm in his travels (Engliſh edition) Vol. II. p. 82 and 311. They hatch only two eggs at a time, and their neſts are built in trees. Their eyes are ſmall and black, the irides yellow, the feet red: the neck finely gloſſed with purple, brighter in the male. They weigh 9 ounces.

V. {Paſſeres.
Paſſerine. Faun. Am. Sept.

8. Alauda. Lark. } 20. Alpeſtris. 289. 10. Klein, Hiſt. of Birds, 4to. p. 73. Shore Lark, Faun. Am. Sept. 12. Cateſb. I. 32.
Albany Fort, N° 6.

This ſpecies is indifferently deſcribed by Linneus, who ſays that all the tail-feathers on their inner web are white, *(rectricibus dimidio interiore albis)*; though it does not appear that he ſaw a ſpecimen of it himſelf. Both the quill

quill and tail-feathers are duſky, and in both the outermoſt feather only has a white exterior margin. The coverts of the tail are of a pale ferruginous colour, and two of them are nearly as long as the tail itſelf. The ſcapulars are ferruginous; in the male, the head and whole back have a tinge of the ſame colour, marked with duſky ſtreaks; in the female, the back is grey, and the duſky ſtripes of a darker hue. The crown of the head is black in the male, duſky in the female; the forehead is yellow, the bill and feet are black, the belly of a dirty reddiſh white. Theſe larks are migratory, they viſit the environs of Albany Fort in the beginning of May, but go further northward to breed: they feed on graſs-ſeeds, and buds of the ſprig-birch; run into ſmall holes, and keep cloſe to the ground, from whence the natives give them the name of *Chi-chup-pi-ſue*.

9. Turdus. } 21. Migratorius, 292. 6. American Thruſh. } Fieldfare. Kalm II. p. 90. Faun. Am. Sept. II. Cateſby I. 29.

Severn River, N° 59. Albany Fort, 7, 8, 9.

The deſcriptions of theſe birds in various authors coincide with the ſpecimens; at Severn River they appear at the beginning of May, and leave the environs before the froſt ſets in. At Mooſe Fort, in the north latitude 51°. they build their neſt, lay their eggs, and hatch their young in the ſpace of fourteen days; but at York fort and Severn ſettlement this is done

done in 26 days: they build their neſts in trees, lay four beautiful light-blue eggs, feed on worms and carrion: when at liberty they ſing very prettily, but confined in a cage, they loſe their melody. There is no material diſtinction between the male and female. Their weight is 2½ ounces, the length 9 inches, and the breadth 1 foot; they are called red birds at Hudſon's Bay; their Indian name is *Pee-pee-chue*.

Turdus, 22.

Severn River, N° 54 and 55, male and female.

From the ſtriking ſimilarity with our blackbird, the Engliſh at Hudſon's Bay have given this bird the ſame name. However, upon a cloſe examination, I find the difference very great between our European blackbird, and the Hudſon's Bay or American one. The plumage of the male, inſtead of being deep black without any gloſs, as in ours, has a ſhining purple caſt, not unlike the plumage of the *Gracula Quiſcula*, Linn. or ſhining Gracule, Faun. Am. Sept.; or the Maize thief, of Kalm. The female indeed is very like our female blackbird, being of a duſky colour on the back, and a dark grey on the breaſt. The feet and bill are quite black in both ſexes; the former have the back claw almoſt as long again as any of the other claws. There are no veſtiges of yellow palpebræ in either the male or the female; the bill in both is ſtrong, ſmooth, and ſubulated; the upper

upper mandible being carinated, but very little arched, and without any tooth or indenture whatever, on the lower ſide. The noſtrils are as in other thruſhes. This bird has no briſtles at the baſe of its bill, its feet have ſuch ſegments as Scopoli in the Annus I. Hiſtorico-Naturalis attributes to the ſtares. Inſtead of being ſolitary and living retired like the European blackbirds, theſe American ones come in flocks to Severn River in June, live among the willows, build in all kinds of trees, and return to the ſouthward in autumn. They feed on worms and maggots; their weight is 2¼ ounces, and they are nine inches long, and one foot broad. One that was kept twelve months in a cage pined away, and died. Notwithſtanding theſe circumſtances, I cannot help remaining undetermined with regard to this bird, which at firſt ſight is like the blackbird, has the bill of a thruſh, and the feet and gregarious nature of a ſtare. It is to be hoped, that future accounts from Hudſon's Bay may inform us further, of the nature of this bird, its time of incubation, the number of eggs it lays, and the colour of thoſe eggs, together with the note of the bird, the difference and characteriſtick marks of both the male and female, and other circumſtances, which may ſerve to determine to what genus and ſpecies we are to refer this bird.

 10. LOXIA

10. LOXIA, Grosbeak. { 23. Curvirostra, 299. 1. Crossbill. Br. Zool. Faun. Am. Sept. 11. The small variety.

Severn River, N° 27 and 28.

This bird comes to Severn River the latter end of May, breeds more to the northward, and returns in autumn, in its way to the south, departing at the setting in of the frost. The irides in the male are of a beautiful red, in the female yellow: the weight is said to be 10 ounces (probably by mistake for 1 ounce, as it is impossible so small a bird should weigh more), the length is 6 inches, the breadth 10.

24. Enucleator, 299. 3. Pine Grosbeak. Br. Zool. and Faun. Am. Sept. Edw. 123, 124. Pl. enl. 135. f. 1.

Severn River, N° 29, 30.

It answers to the descriptions and figures of the ornithologists pretty well; only Edwards's female has the red too bright, which is rather orange in our specimen, on the head, neck, and coverts of the tail. This bird only visits the Hudson's Bay settlements in May, on its way to the north, and is not observed to return in autumn; its food consists of birch-willow buds, and others of the same nature; it weighs 2 ounces, is 9 inches long, and 13 broad.

11. EM-

11. EMBERIZA. { 25. Nivalis. 308. 1. Greater Bunting. { Brambling, Br. Zool. Snowbird Snowflake, ibid. Snow-bunting. Faun. Am. Sept. 11.

Severn River, N° 24—26.

The bird, in ſummer dreſs, correſponds exactly with the deſcription of the greater brambling, Br. Zool. The deſcription of the ſnowflake, or the ſame bird in winter dreſs, ibid. vol. IV p. 19. is ſomewhat different, perhaps owing to the different ſeaſons the birds were caught in, as it is well known they change their colour gradually. They are the firſt of the migratory birds, which come in ſpring to Severn ſettlement; in the year 1771 they appeared April the 11th, ſtayed about a month or five weeks, and then proceeded further northward in order to breed there; they return in September, ſtay till the cold grows ſevere in November, then retire ſouthward to a warmer climate. They live in flocks, feed on graſs-ſeeds, and about the dunghills, are eaſily caught under a ſmall net, ſome oatmeal being ſtrewed under it to allure them; they are very fat, and fine eating. The weight is 1 ounce and 5 drams, the length 6½ inches, and the breadth 10 inches.

EMBRIZA. 26. Leucophrys. *New Species.* White Crowned Bunting.

Severn River, N° 50. Albany Fort, 10.

This elegant little ſpecies of Bunting is called a hedge ſparrow at Hudſon's Bay, and has

 not

not hitherto been defcribed. It vifits Severn fettlement in June, and feeds on grafs-feeds, little worms, grubs, &c. It weighs ¾ of an ounce, and is 7½ inches long, and 9 inches broad; the bill and legs are flefh-coloured; the male is not materially different from the female, its nefts are built in the bottom of willow bufhes, it lays three eggs of a chocolate colour. It vifits Albany Fort in May, breeds there, and leaves it in September.

12. FRINGILLA, Finch. { 27. Lapponica. 317. r. Faun. Suec. 235.

Severn river, N° 52.

It is called *Tecurmafhifh*, by the natives at Hudfon's Bay. The defcription in Linneus's Fauna Suecica coincides exactly with the fpecimen; that in his Syftem anfwers very nearly: Mr. Briffon's defcription (though he quotes Linneus, and Linneus quotes him) is widely different. The fpecimen fent over is a female; the males have more of the ferruginous colour on the head; the eyes are blue, the legs dark brown. It is only a winter inhabitant near Severn river, appears not before November, and is commonly found among the juniper trees; it weighs ½ of an ounce, its length is 5 inches, and its breadth 7.

FRINGILLA.

FRINGILLA. 28. Linaria. 322. 29. Leſſer red headed Linnet. Br. Zool.

Severn River, N° 23.

The deſcriptions of Linneus, Briſſon, and the Britiſh Zoology, anſwer perfectly well. The figure in Planche enluminée 151. f. 2. has a quite ferruginous back contrary to all the deſcriptions and the ſpecimen before us, in which all the feathers on the back are duſky, edged with dirty white.

29. Montana, 324. 37. Mountain Sparrow, Tree Sparrow. Br. Zool. Edw. 269. Briſſon III. p. 79. Faun. Am. Sept.

Severn River, N° 20.

This ſeems to be a variety, as its tail is rather longer than uſual, and forked; it anſwers nearly to the deſcriptions given by the ornithologiſts, and ſeems to be a female, as it has no black under the throat and eyes, and no white collar. The bill and legs are black, the eyes blue. At Severn ſettlement it arrives in May, goes to breed further northwards, and returns in autumn: the weight is ¾ of an ounce, the length 6½ inches, and breadth 10. I was inclined to make this bird a new ſpecies, on account of the many differences between it and the mountain ſparrow; but conſidering the ſpecimen ſent over was not in the beſt order, and might be a female, I thought it beſt to leave it where it is, till we are better informed.

FRIN-

FRINGILLA. 30. Hudſonias. *New Specimen.* Severn River, N° 18.

This is certainly a nondeſcript ſpecies; it only viſits Severn ſettlement in ſummer, not being ſeen there before June, when it ſtays about a fortnight, goes further to the northward to breed, and paſſes by Severn again in autumn on its return ſouth. It is very difficult to procure, and therefore it could not be determined whether the ſpecimen was a male or female. It frequents the plains, and lives on graſs-ſeeds; it weighs ½ an ounce, is 6¼ inches long, and 9 inches broad: it has a ſmall blue eye, and a whitiſh bill faintly tinged with red; the whole body is blackiſh, or of a ſoot colour, the belly alone with the two outermoſt tail feathers on each ſide being white. It is to be wiſhed that more ſpecimens and circumſtantial accounts of this bird were ſent over, which would enable us to determine its character with more preciſion.

13. MUSCICAPA, Flycatcher. { 31. Striata. *New Species,* Striped Flycatcher.

Severn River, N° 48 and 49. Male and Female.

This ſpecies viſits Severn river only in ſummer, feeding on graſs-ſeeds, etc.; it weighs half an ounce, is 5 inches long, and ſeven broad; the male is widely different from the female: this ſpecies is entirely nondeſcript.

14. MOTACILLA, Wagtail. { 32. Calendula. 337. 47. Ruby crowned Wren. Edw. 254. Faun. Am. Sept.

(The number belonging to this bird is loſt; however, it is moſt probably that ſent from Severn river, N° 53.)

It anſwers to the deſcriptions and the figure of Edwards; its weight is 4 drams, its length 4 inches, and its breath 5. It migrates, feeds on graſs-ſeeds and the like, and breeds in the plains; the number of eggs is not known.

15. PARUS, Titmouſe. { 33. Atricapillus. 341. 6. Black Cap Titmouſe.

Albany Fort, N° 11.

The deſcription given by Linneus anſwers, and ſo does M. Briſſon's in moſt particulars, except that the quill-feathers are not white on the inſide. Theſe birds ſtay at Albany Fort all the year, yet ſeem moſt numerous in the coldeſt weather; probably being then more in want of food, they come nearer the ſettlements, in order to pick up all remnants. They feed on flies and ſmall maggots, and likewiſe on the buds of the ſprig-birch, in which they perhaps only ſearch for inſects; they make a twittering noiſe, from which the native call them *Kiſs-kiſs-ke-ſhiſh*.

PARUS. 34. Hudſonicus. *New Species.* Hudſon's Bay Titmouſe.

Severn River, N° 12.

This new ſpecies of titmouſe, is called *Peche-ke-ke-ſhiſh*, by the natives. They are common about the juniper buſhes, of which the buds are their food; in winter they fly about from tree to tree in ſmall flocks, the ſevereſt weather not excepted. They breed about the ſettlements, and lay 5 eggs; they have ſmall eyes, with a white ſtreak under them, and black legs: the male and female are quite alike; they weigh half an ounce, are 5¼ inches long, and 7 inches broad.

16. HIRUNDO, Swallow. } 35.

Severn River, N° 58.

The ſwallows build under the windows, and on the face of ſteep banks of the river, they diſappear in autumn; and the Indians ſay, they were never found torpid under water, probably becauſe they have no large nets to fiſh with under the ice. The ſpecimen ſent anſwers in ſome particulars to the deſcription of the Martin, Hirundo Urbica, Linn. but ſeems to be ſmaller, and has no white on the rump. I have, therefore, thought it beſt to leave the ſpecies undetermined, till further informations are received from Hudſon's Bay, on this ſubject.

2. WATER-

2. WATER-BIRDS.

VI. { GRALLÆ,
Clovenfooted. Faun. Am. Sept.

17. ARDEA, { 36. Canadenſis. 234. 3. Edw. 133.
Heron. { Canada Crane. Faun. Am. Sept. 14.
Severn River, N° 35. Blue Crane.

The account from Severn ſettlement ſays, there is no material difference between the male and female; however, the ſpecimen ſent over, I take to be a female, as its plumage is in general duller than that figured by Edwards, and as the laſt row of white coverts of the wing are wanting. Theſe cranes arrive near Severn in May, have only two young at a time, retire ſouthward in autumn; frequent lakes and ponds, and feed on fiſh, worms, &c. They weigh ſeven pounds and a half, are 3¼ feet long, and 3 feet 5 inches broad; the bill is 4 inches long, the legs 7 inches, but the leg and thigh 19.

ARDEA. 37. Americana, 234. 5. Hooping Crane. Edw. 132. Cateſby, I. 75. Faun. Am. Sept. 14.
York Fort.

Edwards's figure is very exact; Cateſby's is not ſo good, as it repreſents the bill too thick towards the point.

38. Stellaris, 239. 21. *Varietas*. The Bittern, Br. Zool. Edw. 136. Faun. Am. Sept. pag. 14 *. Severn River, N° 64.

At firſt ſight, I thought the ſpecimen ſent from Hudſon's Bay, was a young bird; but upon nearer examination and comparing it with Mr. Edwards's account and figure, I take it to be a variety of the common bittern peculiar to North America; it is ſmaller, but upon the whole very much reſembles our bittern. Mr. Edwards's meaſurements and drawings correſpond very well with the ſpecimen.

This bird appears at Severn river the latter end of May, lives chiefly among the ſwamps and willows, where it builds its neſt, and lays only two eggs at a time; it is very indolent, and, when rouſed, removes only to a ſhort diſtance.

18. SCOLOPAX, Woodcock. { 39. Totanus. 245. 12. Spotted Woodcock. Faun. Am. Sept. 14. Albany Fort, N° 16.

This bird is called a yellow leg at Albany fort, from the bright yellow colour of the legs, eſpecially in old birds; a circumſtance, in which it varies from the deſcriptions of Linneus and Briſſon, probably becauſe they de-

* In the Faunula Americæ Septentrionalis, p. 14. the ſynonym of Ardea Hudſonias, Linn. has by miſtake been annexed to the bittern, and likewiſe pl. 135 of Edwards has been quoted inſtead of plate 136. They are two very different birds.

ſcribed

fcribed from dried fpecimens, in which the yellow colour always changes into brown. It agrees in other refpects perfectly well with the defcriptions: it comes to Albany fort in April or beginning of May, and leaves it the latter end of September. It feeds on fmall fhell fifh, worms, and maggots; and frequents the banks of rivers, fwamps, &c. It is called by the natives *Sa-fa-fhew*, from the noife it makes.

SCOLOPAX. 40. Lapponica. 246. 15. Red Godwit. Br. Zool. Faun. Am. Sept. 14. Ed. 138.
Churchill River, N° 13.

Linneus defcribes this bird very exactly in his Syftema Naturæ: the middle of the belly has no white in the Society's fpecimen, as that had from which the defcription in the Br. Zool. octavo I. p. 353, 354, was taken. All the other characters correfpond.

SCOLOPAX. 41. Borealis. *New Species.* Efkimaux Curlew. Faun. Am. Sept, 14.
Albany Fort, N° 15.

This fpecies of Curlew, is not yet known to the ornithologifts; the firft mention is made of it in the Faunula Americæ Septentrionalis, or catalogue of North American animals. It is called *Wee-kee-me-nafe-fu*, by the natives; feeds on fwamps, worms, grubs, &c; vifits Albany Fort in April or beginning of May; breeds to the northward of it, returns in Au-

 guft,

guſt, and goes away ſouthward again the latter end of September.

19. TRINGA, Sand-piper. { 42. Interpres. 248. 4. Turnſtone. Edw. 141. Faun. Am. Sept. 14. }

Severn River, N° 31 and 32.

This ſpecies is well deſcribed by the ornithologiſts; its weight is 3½ ounces, the length 8¾ inches, and the breadth 17 inches; it has four young at a time; its eyes are black, and the feet of a bright orange: this bird frequents the ſides of the river.

43. Helvetica. 250. 12, Briſſon. Av. V. p. 106. t. 10. f. 2.

(The number was loſt, perhaps it is N° 17, from Fort Albany; upon that ſuppoſition the account is as follows: " the natives call it " *Waw-puſk-abrea-ſhiſh*, or white bear bird; " it feeds on berries, inſects, grubs, worms, " and ſmall ſhell-fiſh; viſits and leaves Al- " bany fort at the ſame time with the *Sco-* " *lopax Totanus*, and *Borealis*.")

I find this bird anſwers very well to its deſcription; the throat, breaſt, and upper part of the belly are blackiſh, as in the deſcriptions, but mixed with white lunulated ſpots, which are neither deſcribed nor expreſſed in M. Briſſon's figure, and may be owing to the difference of ſex, or climate.

VII.

VII. { ANSERES.
Webbed-footed. Faun. Am. Sept.

29. ANAS, { 44. Marila. 196. 8. Scaup Duck. Br.
Duck, { Zool. Faun. Am. Sept. 17.

Severn River, N° 44 and 45. Fiſhing Ducks.

Linneus's deſcription, and the figure in the Br. Zoology, folio, plate Q. p. 153, agree perfectly well with the ſpecimens. The female, as Linneus obſerves, is quite brown, the breaſt and upper part of the back being of a gloſſy reddiſh brown; the ſpeculum of the wing and the belly are white. The eyes of the male have very bright yellow irides; thoſe of the female are of a faint dirty yellow. The female is two ounces heavier than the male, which weighs one pound and an half, is 16½ inches long, and 20 inches broad.

ANAS. 45. Nivalis. Snow Gooſe. Faun. Am. Sept. p. 16. Lawſon's Carolina. Anſer niveus Briſſ. VI. 288. Klein. Anſer nivis. Schwenkfeld, Marſigli. Danub. p. 802. t. 49.

Severn River, N° 40, and a young one, N° 41. white Gooſe.

Theſe white geeſe are very numerous at Hudſon's Bay, many thouſands being annually killed with the gun, for the uſe of the ſettlements. They are uſually ſhot whilſt on the wing, the Indians being very expert at that exerciſe, which they learn from their youth; they weigh five or ſix pounds, are

2⅔ feet

2⅔ feet long, and 3½ broad; their eyes are black, the irides ſmall and red, the legs likewiſe red; they feed along the ſea, and are fine eating; their young are bluiſh grey, and do not attain a perfect whiteneſs till they are a year old. They viſit Severn river firſt in the middle of May, on their journey northward, where they breed; return in the beginning of September, with their young, ſtaying at Severn ſettlement about a fortnight each time. The Indian name is *Way-way*, at Churchill river. Linneus has not taken notice of this ſpecies.

ANAS. 46. Canadenſis. 198. 14. Canada Gooſe. Faun. Am. Sept. 16. Edw. 151. Cateſby I. 92, &c.

Severn River, N° 42.

The Canada geeſe are very plentiful at Hudſon's Bay, great quantities of them are ſalted, but they have a fiſhy taſte. The ſpecimen ſent over agrees perfectly with the deſcriptions and drawings. At Hudſon's Bay this ſpecies is called the *Small Grey Gooſe*. Beſides this, and the preceding white gooſe, Mr. Graham, the gentleman who ſent the account from Severn ſettlement, mentions three other ſpecies of wild geeſe to be met with at Hudſon's Bay; he calls them,

1. The large Grey Gooſe.
2. The Blue Gooſe.
3. The Laughing Gooſe.

The first of these, the large grey goose, he says, is so common in England, that he thought it unnecessary to send specimens of it over. It is however presumed, that though Mr. Graham has shewn himself a careful observer, and an indefatigable collector; yet, not being a naturalist, he could not enter into any minute examination about the species to which each goose belongs, nor from mere recollection know, that his grey goose was actually to be met with in England. A natural historian, by examination, often finds material differences, which would escape a person unacquainted with natural history. The wish, therefore, of seeing the specimens of these species of geese, must occur to every lover of that science. Mr. Graham says, the large grey geese are the only species that breed about Severn river. They frequent the plains and swamps along the coast. Their weight is nine pounds.

The blue goose is as big as the white goose; and the laughing goose is of the size of the Canada or small grey goose. These two last species are very common along Hudson's Bay to the southward, but very rare to the northward of Severn river. The Indians have a peculiar method of killing all these species of geese, and likewise swans. As these birds fly regularly along the marshes, the Indians range themselves in a line across the marsh, from the wood to high water mark, about musket shot from eath other, ſo as to be ſure of intercepting any geeſe which fly that way. Each perſon conceals himſelf, by putting round him ſome bruſh wood; they likewiſe make artificial geeſe of ſticks and mud, placing them at a ſhort diſtance from themſelves, in order to decoy the real geeſe within ſhot: thus prepared, they ſit down, and keep a good look out; and as ſoon as the flock approaches, they all lie down, imitating the call or note of geeſe, which theſe birds no ſooner hear, and perceive the decoys, than they go ſtraight down towards them; then the Indians riſe on their knees, and diſcharge one, two or three guns each, killing two or even three geeſe at each ſhot, for they are very expert. Mr. Graham ſays, he has ſeen a row of Indians, by calling round a flock of geeſe, keep them hovering among them, till every one of the geeſe was killed. Every ſpecies of geeſe has its peculiar note or call, which muſt greatly increaſe the difficulty of enticing them.

ANAS. 47. Albeola. 199. 18. The Red Duck. Faun. Am. Sept. 17. Edw. t. 100. Sarcelle de la Louiſiane. Briſſon VI. t. 41. f. 1.

Severn River, N° 37 and 38. Fiſhing Birds.

The deſcriptions and figures anſwer very well with the male, except that the three exterior feathers are not white on the outſide, but all duſky.

The female is not deſcribed by any one of the ornithologiſts; and therefore deſerves to be noticed,

noticed, to prevent future miſtakes. The whole bird is duſky, a few feathers on the forehead are ruſty, and ſome about the ears of a dirty white; the breaſt is grey, the belly and ſpeculum in the wings white; the bill and legs are black. They viſit Severn ſettlement in June, build their neſts in trees, and breed among the woods, and near ponds; the weight of the female is one pound, its length 14 inches, and its breath 21.

ANAS. 48. Clangula. 201. 23. Golden Eye. Br. Zool. Faun. Am. Sept. 16.

Severn River, N° 51.

Theſe birds frequent lakes and ponds, and breed there: they eat fiſh and ſlime, and cannot riſe off the dry land. The legs and irides are yellow; their weight is 2 1/8 pounds, and their meaſure 19 inches in length, and two feet in breadth. The ſpecimen ſent is the male.

ANAS. 49. Perſpicillata. 201. 25. Black Duck. Faun. Am. Sept. 16. Edw. 155.

Churchill River, N° 14.

This ſpecies is exactly deſcribed, and well drawn by Edwards. The Indians call it *She-ke-ſu-partem.* It ought to come into the firſt diviſion of Linneus's ducks, "roſtro baſi "gibbo," as its bill is really very unequal at the baſe.

ANAS. 50. Glacialis. 203. 30, and Hyemalis, 202. 29. Edw. t. 156. Swallow-tail. Br. Zool. Faun. Am. Sept. 17.

Churchill River, N° 12.

At Churchill River the Indians call this ſpecies, *Har-har-vey*; it correſponds with Edwards's deſcription and drawing, plate 156, but diſfers much from Linneus's inexact deſcription of the Anas Hyemalis, to which he, however, quotes Edwards. Upon the whole it is almoſt without a doubt that the bird repreſented by Edwards, plate 280, and Br. Zool. folio, plate Q. 7, and quoted by Linneus for his Anas glacialis, is the male, and that the bird figured by Edwards t. 156, and quoted by Linneus for the Anas Hyemalis, is the female, of one and the ſame ſpecies. Linneus mentions a white body (in his Anas hyemalis) which in Edw. Tab. 156, and in the Society's ſpecimen, is all brown and duſky, except the belly, temples, a ſpot on the back of the head, and the ſides of the rump, which are white. Linneus ſays, that the temples are black; in the ſpecimen now ſent over, and in Mr. Edwards's figure, which Linneus quotes, they are white; the breaſt, back, and wings, are not black as he ſays, but rather brown and duſky. A further proof, that Linneus's Anas Glacialis and Hyemalis are the ſame, is that the feet in both t. 156 and 280 of Edwards are red, and the bill black, with an orange ſpot.

ANAS. 51. Crecca. 204. 33. *Varietas.* Teal. Br. Zool. Faun. Am. Sept. 17.

Severn River, N° 33, 34. Male and female.

This is a variety of the teal, for it wants the two white ſtreaks above and below the eyes; the lower one indeed is faintly expreſſed in the male, which has alſo a lunated bar of white over each ſhoulder; this is not to be found in the European teal. This ſpecies is not very plentiful near Severn river; they live in the woods and plains near little ponds of water, and have from five to ſeven young at a time.

ANAS. 52. Hiſtrionica. 204. 35. Harlequin Duck. Faun. Am. Sept. 16. Edw. t. 99.

This bird had no number fixed to it; it agrees perfectly with Edwards's figure.

ANAS. 53. Boſchas. 205. 40. Mallard Drake. Faun. Am. Sept. Br. Zool.

Severn River, N° 39.

It is called Stock Drake at Hudſon's Bay, and correſponds in every reſpect with the European one, upon compariſon.

21. PELECANUS, Pelecan. } 54. Onocrotalus. 251. 1. *A variety.*

York Fort.

This variety of the pelecan, agrees in every paticular with Linneus's oriental pelecan (Pelecanus onocrotalus orientalis), but has a peculiar tuft or fringe of fibres in the middle of the upper mandible, ſomething nearer the apex than the baſe. This tuft has not been mentioned by any author, and is likewiſe wanting in Edwards's pelican, t. 92. with which the Society's ſpecimen correſponds in every other circumſtance. The P. Onocrotalus occidentalis, Linn. or Edw. t. 93 American pelican, is very different from it: the chief differences are the colour, which in our Hudſon's Bay bird is white, but in Edwards's is of a greyiſh brown; and the ſize, which in the white bird is almoſt double of the brown one. The quill-feathers are black, and the ſhafts of the larger ones white. The *Alula*, or baſtard wing, is black. The bill and legs are yellow.

22. COLYMBUS. * Diver, } 55. Glacialis. 221. 5. Northern Diver. Br. Zool. Faun. Am. Sept. 16.

Churchill River, N° 8. called a Loon there.

This bird is well deſcribed and drawn in the Britiſh Zoology, in folio.

* * Grebe. } 56. Auritus, α. 222. 8. Edw. 145. Eared Grebe. Faun. Am. Sept. 15.

Severn River, N° 43.

This is exactly the bird drawn by Edwards, t. 145. The ſpecimen ſent over is a female. It differs much from our leſſer creſted Grebe. Br.

Br. Zool. octavo I. p. 396, and Br. Zool. illustr. plate 77. fig. 2. and Ed. 96. fig. 2. However, in both these works, it is looked on only as a variety, or different in sex. Mr. Graham has the same opinion. It lives on fish, frequenting the lakes near the sea coast. It lays its eggs in water, and cannot rise off dry land. It is seen about the beginning of June, but migrates southward in autumn. It is called *Sekeep*, by the natives. Its eyes are small, the irides red; it weighs one pound, and measures one foot in length, and one third more in breadth.

23. LARUS. Gull. } 57. Parasiticus. 226. 10. Arctic Gull. Br. Zool. Faun. Am. Sept. 16. Edw. 148. 149.

Churchill River, N° 15.

This species is called a *Man of War*, at Hudson's Bay. It seems to be a female, by the dirty white colour of its plumage below; it agrees very well with Edwards's drawing, and that in the Br. Zool. illustr.

24. STERNA. Tern. } 58. Hirundo *(Variety)*, 227. 2. The greater Tern. Br. Zool. Faun. Am. Sept.

(The number belonging to this bird is lost, perhaps it is N° 17, from Churchill River, called

" A sort

" A ſort of Gull, called Egg-breakers, by " the natives.")

The feet are black; the tail is ſhorter and much leſs forked than that deſcribed and drawn in the Br. Zool. The outermoſt tail-feather likewiſe wants the black, which that in the Britiſh Zoology has. In other reſpects it is the ſame.

DESCRIP-

DESCRIPTIONES Avium Rariorum e Sinu Hudſonis.

1. Falco sacer.

Falco, cerâ pedibuſque coeruleis, corpore, remigibus rectricibuſque fuſcis, faſciis pallidis; capite, pectore & abdomine albis, maculis longitudinalibus fuſcis.

Habitat ad ſinum Hudſonis et in reliqua America Septentrionali; victitat Lagopodibus & Tetraonum ſpeciebus.

Descr. *Magnitudo* Corvi.

Roſtrum, cera, pedes coerulea; roſtrum breve, curvum, coeruleo-atrum; mandibula utraque, baſi pallide coerulea, apice nigreſcente, utraque emarginata.

Caput tectum pennis albidis, maculis longitudinalibus, fuſcis.

Oculi magni; irides flavæ.

Gula alba, fuſco-maculata.

Dorſum et tectrices alarum, plumis fuſcis, ferrugineo-pallide marginatis, maculatiſque, maculis rachin non attingentibus.

Pectus, venter, criſſum, tectrices alarum inferiores, & femora alba, maculis longitudinalibus nigro-fuſcis.

Remiges fuſco-nigri, viginti duo; primores apicibus margine albis, maculis ferrugineo-pallidis, intra majoribus, transversis, extra minoribus, rotundatis.

Rectrices duodecim, supra fuscæ, fasciis circiter duodecim & apice albidis; infra cinereæ, fasciis albidis.

2. STRIX NEBULOSA.

STRIX capite lævi, corpore fusco, albido undulatim striato, remige sexto longiore, apice nigricante.

Habitat circa Sinum Hudsonis, victitat Leporibus, Lagopodibus, Muribusque.

DESCR. *Rostrum* fusco-flavum, mandibula superiore superius magis flava.

Oculi magni, iridibus flavis.

Caput facie cinerea, e pennis fusco et pallide cinereo alternatim striatis. Pone hasce pennas collum versus est ordo plumularum fuscarum ad utramque genam, semicirculum nigrum efficiens.

Occiput, cervix, et collum fusca, pennis, marginibus albo-maculatis.

Pectus albidum, maculis longitudinalibus transversisque fuscis.

Abdomen album, superius uti pectus maculis longitudinalibus, sed inferius striis transversis notatum.

Dorsum totum et tectrices alæ, caudæque confertim ex fusco & albido undulato-striatæ.

Alæ fuscæ; remiges primores fusci, griseo transversim fasciati, fasciis latis nebulosis. Remex sextus, reliquis longior, apice

 magis

magis nigricans; primus vero reliquis primoribus brevior. Remiges reliqui pallidiores, obſcurius faſciati.

Cauda rotundata, rectricibus duodecim: duæ intermediæ paullo longiores, totæ cineraſcente albido fuſcoque undulatim ſtriatæ, lineis duplicatis fuſcis tranſverſis pluribus. Rectrices reliquæ fuſcæ albido ſubſtriatæ.

Pedes tecti pennis albidis fuſco-ſtriatis.

Magnitudo fere Strigis Nycteæ, Linn.

Longitudo unciarum 16 pedis Anglicani.

Latitudo pedum quatuor.

Pondus librarum trium.

3. TETRAO PHASIANELLUS.

Linn. Ed. X. p. 160. n. 5.

TETRAO pedibus hirſutis, cauda cuneiformi, remigibus nigris, exterius albo-maculatis.

Habitat ad Sinum Hudſonis.

DESCR. *Magnitudo* fere Tetraonis Tetricis. Linn.

Roſtrum nigrum.

Oculorum irides avellaneæ.

Caput, collum & dorſum teſtacea, nigro tranſverſim faſciata: macula albida inter roſtrum et oculos: latera colli notata maculis rotundatis albidis.

Dorſum teſtaceum, plumis omnibus late nigro-faſciatis.

Uropygium magis albido-cinereum, nigredine fimbriata ſecundum rachin plumarum.

Pectus & Venter albida, maculis cordatis fuſco-teſtaceis in ventre ſaturatioribus.

Alarum tectrices dilute teſtaceo, nigro, alboque tranſverſim faſciatæ, maculis pluribus rotundis albis. *Remiges* primores nigri, latere exteriore albo-maculati; ſecundarii fuſci, apice & ad marginem exteriorem albo ſubfaſciati: poſtremi vero teſtaceo faſciati, apice tantum albi.

Rectrices breves, exteriores pallide fuſcæ, apice albæ, duæ intermediæ reliquis longiores, teſtaceo-maculatæ.

Pedes plumis albo-griſeis veſti digitis pectinatis.

Longitudo unciarum 16 pedis Anglicani.

Latitudo pedum duorum.

4. Emberiza leucophrys *.

Emberiza remigibus rectricibuſque fuſcis, capite nigro, faſcia verticis, ſuperciliiſque niveis.

Habitat in America Boreali ad Sinum Hudſonis.

Descr. *Magnitudo* circiter *fringillæ cælibis.*

Roſtrum rubrum, ſ. carnei coloris: Nares ſubrotundæ.

Caput faſcia verticali lata candida, paululum ante roſtrum deſinente; faſcia atra

* Λευκὸς albus. Ὀφρὺς ſupercilium.

lata ad utrumque latus fasciæ albæ. Supercilia alba, desinentia in lineas, fasciam albam verticalem adtingentes; arcus dein atri, ex angulis oculorum, fere in occipite confluentes.

Collum cinerascens, in pectore dilutius.

Dorsum ferrugineo-fuscum, marginibus plumularum cinereis.

Alæ fuscæ; remigum primorum margines exteriores tenuissimi pallidi, interiores cinerascentes: secundarii & pennæ tectrices fuscæ, marginibus latiusculis, versus apicem albis, efficientibus fasciam albam; super quam fascia altera alba ex maculis albis in apice tectricum minorum, s. plumarum scapularium. Alulæ albæ. Remiges subtus cinerei, marginibus albis.

Pectus cinereum, abdomen dilutius, fere album.

Crissum & plumulæ femora tegentes lutescentia.

Uropygium cinereo-fuscum.

Cauda æqualis; rectrices duodecim fuscæ, marginibus paullo pallidioribus, subtus cinereæ.

Pedes carnei coloris, digito intermedio & ungue postico reliquis longioribus.

Longitudo unciarum 7 pedis Anglicani.

Latitudo inter alas extensas 9 unciarum pedis Anglicani.

Cauda partem tertiam longitudinis totius aviculæ efficit.

Alæ complicatæ paulululum ultra caudæ exortum protenduntur.
Pondus drachmarum ſex.

5. FRINGILLA HUDSONIAS.

FRINGILLA fuſco-cineraſcens, roſtro albido, pectore inferiore, abdomine, rectricibuſque quatuor extremis albis.

Habitat in America Boreali.

DESCR. *Magnitudo* circiter fringillæ carduelis.
Roſtrum albidum, rubedine aliqua imbutum.
Oculi parvi, cœrulei.
Corpus totum cinereo-nigricans, ſ. potius fuliginoſum.
Pectus inferius & *abdomen* alba.
Remiges fuſci, cinereo-marginati: alæ complicatæ mediam fere caudam adtingunt.
Rectrices fuſcæ, extimæ utrinque duæ totæ albæ, tertia fuſca, macula oblonga alba, ad latus interius, prope rachin, apicem attingens; reliquæ totæ fuſcæ.
Pondus ſemunciæ.
Longitudo unciarum 6¼ pedis Anglicani.
Latitudo unciarum novem.

6. MUSCICAPA STRIATA.

MUSCICAPA cinereo-virens, dorſo nigro ſtriato, ſubtus flaveſcenti-alba, gula lateribuſque pectoris fuſco maculatis.

Habitat

Habitat ad Sinum Hudſonis.

Quum mas à fœmina multum differat, utique congruum eſt, utrumque ſexum ſeparatim deſcribere.

DESCR. Mas.

Roſtrum trigonum, mandibu ſuperiore paululum longiore, ante apicem leviter emarginata, nigra; inferiore baſi flaveſcente.

Nares ſubrotundæ.

Vibriſſæ nigræ.

Caput ſupra totum atrum ad oculos uſque. *Genæ* à roſtro in occiput totæ albæ; occiput albo & nigro variegatum.

Gula flaveſcenti-alba maculis fuſcis.

Pectus albidum, lateribus, ſive verſus occiput maculis nigris variegatum.

Dorſum cinereo-virens, ſtriis ſive maculis longitudinalibus nigris latioribus, è plumulis nigris, margine virentibus.

Abdomen album.

Uropygium cinereum, nigro-maculatum.

Alæ fuſcæ; remiges primores pallido marginati, ſecundarii apice tenuiſſimo albo; duæ ultimæ margine exteriore albo; tectrices fuſcæ, majores flaveſcenti albo, minores candido in apice maculatæ, unde faſciæ albæ binæ in alis.

Cauda fuſca; rectrix utrinque prima ſ. extima, latere interiore macula magna alba, marginem interiorem attingente; proxima ſ. ſecunda macula oblonga minore alba, etiam marginem interiorem attingente;

attingente; utrinque tertia, latere interiore verſus apicem albo-marginata.

Pedes lutei; ungues breves, pallide fuſci.

Magnitudo circiter *Pari atricapilli*; Linn.

Longitudo 5 unciarum.

Latitudo 7 unciarum pedis Anglicani.

Fœmina.

Roſtrum, alæ, cauda, abdomen, uropygium, pedes & menſuræ ut in mare.

Caput flavo-virens, ſtriis brevibus tenuibuſque longitudinalibus nigris; linea flaviſſima à baſi roſtri incipiens ſuper oculos ducta; palpebræ flavæ.

Gula, genæ & pectus albido-flava; maculæ ſparſæ oblongiuſculæ fuſcæ, ab utroque oris angulo uſque in pectoris latera.

Dorſum, ut in mare, ſed viridius, & ſtriæ nigræ minores.

7. Parus Hudsonicus.

Parus capite fuſco-rubeſcente, dorſo cinereo, jugulo atro, faſcia ſuboculari, pectoreque albis, hypochondriis rufis.

Habitat ad Sinum Hudſonis.

Descr. *Roſtrum* ſubulatum, integerrimum, atrum, baſi è regione narium tectum faſciculis ſetarum ferruginearum, lineas 4 (unciæ pedis Anglicani) longum.

Caput fuſco-ferrugineum, faſcia ſub oculis alba; gula atra, nigredine extenſa ſub hac faſcia alba.

Dorſum

Dorſum cinereo-virens, è plumis longioribus, fuſcis, apice tantum cinereo-virentibus, ſ. olivaceis.

Pectus & Abdomen alba, ſed plumæ omnes baſi nigræ, apice tantum albæ.

Latera abdominis & lumbi ferruginei.

Alæ fuſcæ, remigum margine omni cinereo.

Cauda fuſca, rotundata, rectricibus 12, margine cinereis.

Uropygium tectum plumulis aliquot nigris, apice albido-rufis.

Pedes nigri; digitus poſticus cum ungue anticorum digitorum medio, duplo longior.

Longitudo unciarum $5\frac{1}{8}$ pedis Anglicani.

Latitudo unciarum 7.

Cauda uncias $2\frac{1}{2}$ longa.

8. SCOLOPAX BOREALIS.

SCOLOPAX roſtro arcuato, pedibuſque nigris, corpore fuſco, griſeo-maculato, ſubtus ochroleuco.

Habitat in Sinus Hudſonis inundatis, & pratis humidis, victitans vermibus & inſectis: menſe Aprili vel initio Maii primum viſa eſt, circa Caſtellum *Albany*, inde in terras magis arcticas migrat, ibique nidificat; redit ad idem caſtellum menſe Auguſto; regiones Auſtraliores petit circa finem Septembris.

Affinis ſcolopace arquata Linn. ſed differt corpore triplo minore, roſtro ratione corporis breviore,

breviore, colore in dorſo ſaturate fuſco, in abdomine ochroleuco.

DESCR. *Caput* pallidum, lineolis confertis longitudinalibus fuſcis: ſinciput ſaturate fuſcum, pallido maculatum.

Roſtrum nigricans, arcuatum, longitudine duarum unciarum pedis Anglicani, mandibula inferiore baſi rufa.

Collum, pectus, abdomen & criſſum ochroleuca; pectore colloque lineolis longitudinalibus fuſcis confertioribus, abdomine & criſſo fere nullis, vel tenuibus notatis.

Femora ſemi-tecta plumulis ochroleucis, fuſco maculatis.

Latera abdominis ſub alis præſertim, rufa, pennis tranſverſim fuſco faſciatis.

Dorſum totum ſaturate fuſcum, pennis margine albido griſeis.

Alæ fuſcæ; remiges primores immaculati, primores rachi tota alba; reliqui, ſ. ſecundarii pallide griſeo-marginati. Tectrices late griſeo-marginatæ. Tectrices inferiores alæ, ferrugineæ fuſco tranſverſim faſciatæ. Alæ complicatæ fere mediam caudam attingunt.

Uropygium fuſcum, marginibus maculiſque pennarum albidis.

Cauda brevis, fuſca, rectricibus albido tranſverſim faſciatis

Pedes nigri, ſ. cœruleſcentes.

Longitudo unciarum 13½.

Latitudo circiter unciarum 21.

9. ANAS NIVALIS.

ANAS, roſtro cylindrico, corpore albo, remigibus primoribus nigris.

Habitat in America Boreali, per Sinum Hudſonis migrans.

DESCR. *Corpus* totum album, magnitudine anſeris domeſtici noſtratis.

Roſtrum luteum, mandibulis ſubſerratis.

Oculi iride rubra.

Remiges decem primores nigri, ſcapis albis: tectrices infimæ cinereæ, ſcapis nigris; pennæ duæ alulæ, itidem cinereæ, ſcapis nigris.

Pedes rubri.

Longitudo pedum duorum & unciarum octo.

Latitudo pedum 3½.

Pondus librarum 5 vel 6.

XXX. *Geometrical Solutions of three celebrated Aſtronomical Problems, by the late Dr.* Henry Pemberton, *F. R. S. Communicated by* Matthew Raper, *Eſq; F. R. S.*

LEMMA.

Read June 4, 1772.

TO form a triangle with two given ſides, that the rectangle under the ſine of the angle contained by the two given ſides, and the tangent of the angle oppoſite to the leſſer of the given ſides, ſhall be the greateſt that can be.

Let [TAB. XII. Fig. 1.] the two given ſides be equal to A B and A C: round the center A, with the interval A C, deſcribe the circle C D E, and produce B A to E; take B F a mean proportional between B E and B C, and erect the perpendicular F G, and complete the triangle A G B.

Here the ſine of B A G is to the radius, as F G to A G; and the tangent of A B G to the radius, as F G to F B: therefore, the rectangle under the ſine of B A G and the tangent of A B G is to the ſquare of the

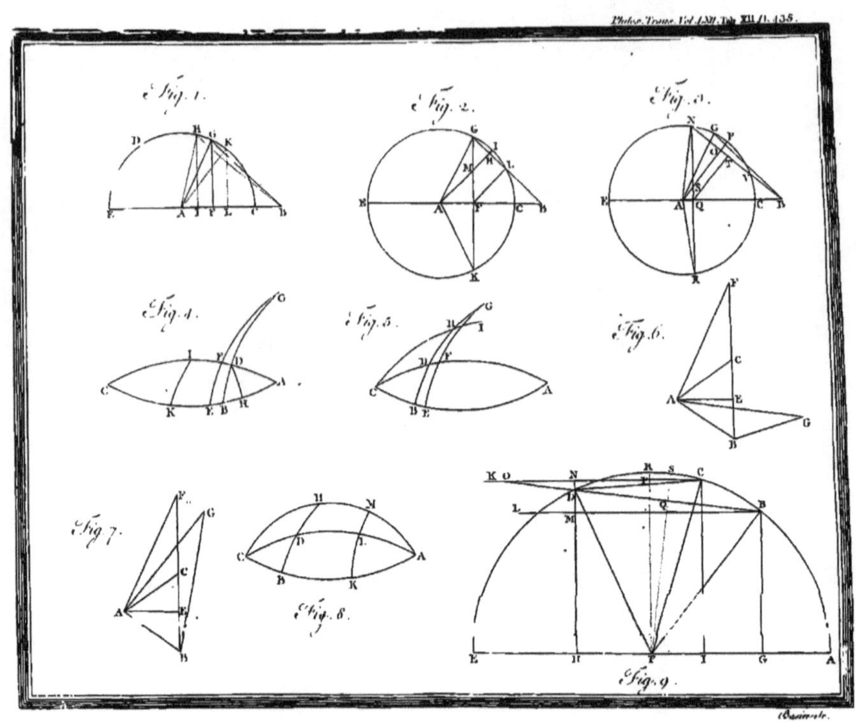
Philos. Trans. Vol. LXII. Tab. XII. p. 435.
Fig. 1.
Fig. 2.
Fig. 3.
Fig. 4.
Fig. 5.
Fig. 6.
Fig. 7.
Fig. 8.
Fig. 9.
Basire sc.

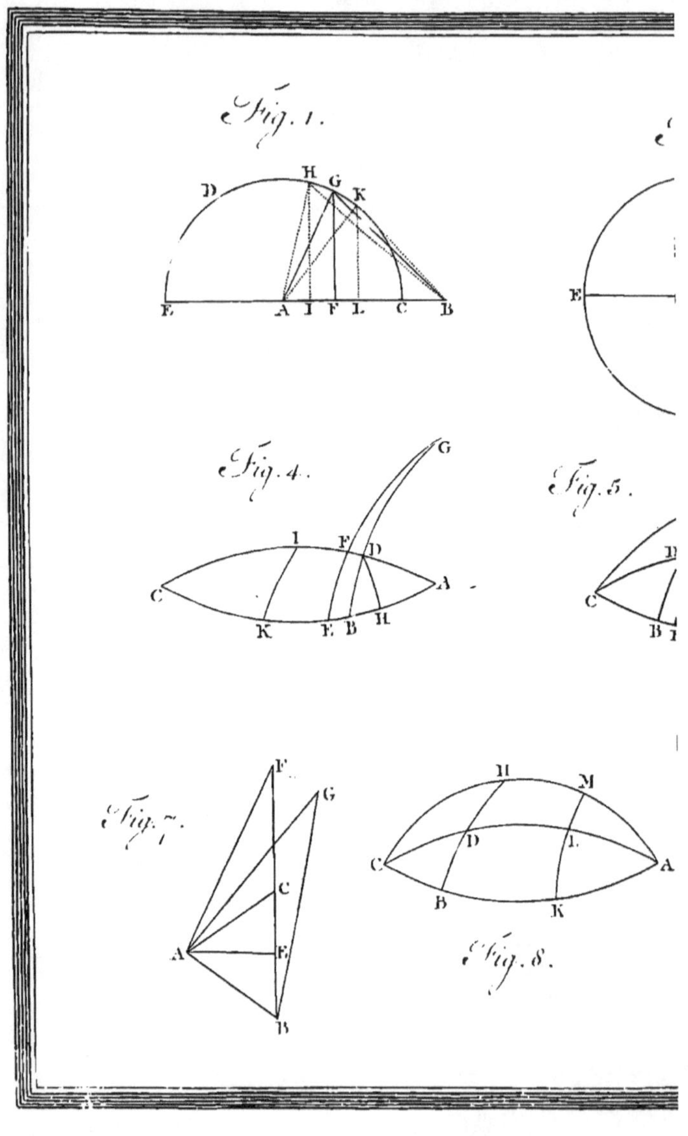
Fig. 1.
H G K
D
E A I F L C B
E
Fig. 4.
G
I F D
C A
K E B H
Fig. 5.
C
B
Fig. 7.
F
G
C
A E
B
H M
D L
C A
B K
Fig. 8.

the radius, as the ſquare of F G, or the rectangle E F C, to the rectangle under A G (or A C) and F B, But, E B being to B F as B F to B C, by converſion, E B is to E F as B F to F C, and alſo, by taking the difference of the antecedents and of the conſequents, E F is to twice A F as B F to F C; and twice A F B is equal to E F C.

Now, let the triangle B A H be formed, where the angle B A H is greater than B A G. Here, the perpendicular H I being drawn, the rectangle under the ſine of B A H and the tangent of A B H will be to the ſquare of the radius, as the rectangle E I C to the rectangle under A C, I B. But I F is to F B as 2 A F I to 2 A F B, or E F C; and 2 A F I is greater than $AF^q - AI^q$; alſo $AF^q - AI^q$ together with E F C, is equal to E I C; therefore, by compoſition, the ratio of I B to B F is greater than that of E I C to E F C; and the ratio of $AC \times IB$ to $AC \times FB$ greater than that of E I C to E F C: alſo, by permutation, the ratio of $AC \times IB$ to E I C greater than the ratio of $AC \times FB$ to E F C. But the firſt of theſe ratios is the ſame with that of the ſquare of the radius to the rectangle under the ſine of B A H and the tangent of A B H; and the latter is the ſame with that of the ſquare of the radius to the rectangle under the ſine of B A G and the tangent of A B G; therefore, the latter of theſe two rectangles is greater than the other.

Again, let the triangle B A K be formed, with the angle B A K leſs than B A G, and the perpendicular K L be drawn. Then the rectangle under the ſine of B A K and the tangent of A B K is to the ſquare of the radius, as the ſquare of K L to the rectangle under

 A C,

AC, BL. Here, FL being to FB as 2AFL to 2AFB or EFC, and 2AFL less than $AL^2 - AF^2$, by conversion, the ratio of LB to FB will be greater than the ratio of ELC to EFC; therefore, as before, the rectangle under the sine of BAG and the tangent of ABG is greater than that under the sine of BAK and the tangent of ABK.

Corollary 1.

BF is equal to the tangent of the circle from the point B; therefore, BF is the tangent, and AB the secant, to the radius AC, of the angle, whose cosine is to the radius as AC to AB. Therefore, AF is the tangent, to the same radius, of half the complement of that angle; and AF is also the cosine of the angle BAG to this radius.

Corol. 2.

The sine of the angle composed of the complement of AGB, and twice the complement of ABG, is equal to three times the sine of the complement of AGB. Let fall the perpendicular AH (Fig. 2.), cutting the circle in I; continue GF to K, and draw AK. Then $BF^2 = EBC = GBL$. Therefore, GB : BF :: BF : BL, and the triangles GBF, FBL are similar. Consequently FL is perpendicular to GB, and parallel to AH; whence GH being equal to HL, GM is equal to MF, and MK equal to three times GM.

Now, the arc IK = 2IC + GI; and the angle IAK = 2IAC + GAI; also GM is to MK as the

the ſine of the arc GI to the ſine of the arc IK, that is, as the ſine of the angle GAI to the ſine of the angle IAK. Therefore, the ſine of the angle IAK (=2IAC + GAI) is equal to three times the ſine of the angle GAI; but GAI is the complement of AGB, and IAC the complement of ABG.

Corol. 3.

If (Fig. 3.) any line BN be drawn to divide the angle ABG, and AN be joined, alſo AO be drawn perpendicular to BN, and continued to the circle in P, the ſine of the angle compoſed of NAP and 2PAC will be leſs than three times the ſine of the angle NAP. Draw NQR perpendicular to AB, cutting AP in S; join AR, and draw QT perpendicular to BN, and parallel to AO; then $BQ^q = NBT$. But BQ^q is greater than the rectangle EBC, that is, greater than the rectangle NBV, under the two ſegments of the line BN drawn from B, to cut the circle in N and V: therefore, TB is greater than VB, and NO greater than OT. Conſequently NS is greater than SQ. Hence RS is leſs than three times NS; and therefore, the ſine of the angle PAR (=NAP+2PAC) is leſs than three times the ſine of NAP.

Problem

PROBLEM I.

To find in the ecliptic the point of longeſt *aſcenſion.*

ANALYSIS.

Let (Fig. 4.) ABC be the equator, ADC the ecliptic, BD the ſituation of the horizon, when D is the point of longeſt aſcenſion. Let EFG be another ſituation of the horizon. Then the ratio of the ſine of EB to the ſine of FD is compounded of the ratio of the ſine of EG to the ſine of GD, and of the ratio of the ſine of AE to the ſine of AF; but the angles B and E being equal, the arcs EG, GB together make a ſemicircle; and, by the approach of EG towards GB, the ultimate magnitude of BG will be a quadrant, and the ultimate ratio of EB to FD will be compounded of the ratio of the radius to the ſine of DG (that is, the coſine of BD) and of the ratio of the ſine of AB to the ſine of AD. Draw the arc DH perpendicular to AB. Then, in the triangle BDH, the radius is to the coſine of BD, as the tangent of the angle BDH to the cotangent of HBD. Alſo, in the triangle BDA, the ſine of AB is to the ſine of AD as the ſine of the angle BDA (or BDC) to the ſine of ABD; therefore, the ultimate ratio of BE to DF is compounded of the ratio of the tangent of BDH to the cotangent of ABD, and of the ratio of the ſine of BDC to the ſine of ABD; which two ratios compound that of the rectangle under the tangent of BDH and the ſine of BDC to the rectangle under the cotangent and the ſine of the given angle ABD.

But, when D is the point of longeſt aſcenſion, the ratio of BE to DF is the greateſt that can be; therefore, then the ratio of the rectangle under the tangent of BDH and the ſine of BDC to the given rectangle under the cotangent and ſine of the given angle ABD muſt be the greateſt that can be; and conſequently, the rectangle under the tangent of BDH, and the ſine of BDC, muſt be the greateſt that can be.

In the triangle BDA, the ſine of BDH is to the ſine of HDA, as the coſine of ABD to the coſine of BAD. Now, in the preceding lemma, let the angle BAG of the triangle AGB be equal to the ſpherical angle BDC: then will the ſum of the angles ABG, AGB be equal to the ſpherical angle BDA. And, if AG in the triangle AGB, be to AB as the coſine of the ſpherical angle DBA to the coſine of DAB, that is, as the ſine of BDH to the ſine of HDA, the angle ABG, in the triangle, will be equal to the ſpherical angle BDH; and the angle AGB, in the triangle, equal to the ſpherical angle HDA. Therefore, by the firſt corollary of the lemma, that the rectangle under the tangent of the ſpherical angle BDH and the ſine of BDC be the greateſt that can be, the coſine of BDC muſt be equal to the tangent of half the complement of the angle, whoſe coſine is to the radius, as AG to AB, in the triangle, or as the coſine of the ſpherical angle ABD to the coſine of the ſpherical angle BAD.

If IK be the ſituation of the horizon, when the ſolſtitial point is aſcending, in the quadrantal triangle AIK, the coſine of KIC is to the radius as the coſine of IKA (= DBA) to the coſine of IAK. Therefore,

fore, the cosine of BDC, when D is the point of longest ascension, is equal to the tangent of half the complement of the angle, which the ecliptic makes with the horizon, when the solstitial point is ascending.

But, the sine of the angle composed of DAB, and twice ABD, must be less than three times the sine of the angle BAD. In the spherical triangle ABD, the angles BAD, ABD together exceed the external angle BDC. Therefore, in the third corollary of the lemma, let the angle BAN be equal to the sum of the spherical angles BAD, ABD: but here, AN is to AB as the cosine of the spherical angle ABD to the cosine of BAD; and AN is also to AB as the sine of ABN to the sine of ANB, that is, as the cosine of BAP to the cosine of NAP; consequently, since the angle BAN is equal to the sum of the spherical angles BAD, ABD, the angle NAP is equal to the spherical angle BAD, and the angle BAP equal to the spherical angle ABD; but the sine of the angle composed of NAP and twice PAB is less than three times the sine of NAP; therefore, the sine of the angle composed of the spherical angle BAD and 2 ABD will be less than three times the sine of the angle BAD; otherwise no such triangle DBA, as is here required, can take place, but the point A will be the point of longest ascension.

If the sine of the angle A be greater than one third of the radius, the point A can never be the point of longest ascension; but when the sine of this angle is less, the angle compounded of BAD and twice ABD, may be greater or less than a quadrant;

and

and therefore, the magnitude of the angle ABD, that A be the point of longeſt aſcenſion, is confined within two limits, of which the double of one added to the angle A, as much exceeds a quadrant, as the double of the other added to that angle falls ſhort of it; therefore, double the ſum of thoſe two angles, together with twice A, makes a ſemicircle; and the ſingle ſum of thoſe two angles added to A makes a quadrant.

PROBLEM II.

To find when the arc of the ecliptic differs moſt from its oblique aſcenſion.

ANALYSIS.

If (Fig. 5.) BD be the ſituation of the horizon, when CD differs moſt from CB, as before, the ultimate ratio of BE to DF will be compounded of the ratio of the radius to the ſine of DG (or the coſine of DB) and of the ratio of the ſine of CB to the ſine of CD: but, when CD differs moſt from CB, BE and DF are ultimately equal; therefore, then the coſine of BD is to the radius as the ſine of CB to the ſine of CD.

Draw the arc CHI of a great circle, that DH be equal to DB; then, BH being double BD, half the ſine of BH is to the ſine of BD or DH, as the coſine of BD to the radius; therefore, half the ſine of BH is to the ſine of DH as the ſine of CB to the ſine of CD; but the ſine of the angle BCH is to the ſine of BH as the ſine of the angle CHB to the

 ſine

ſine of CB; whence, by equality, half the ſine of BCH is to the ſine of DH as the ſine of CHB to the ſine of CD: but as the ſine of CHB to the ſine of CD, ſo, in the triangle CHD, is the ſine of DCH to the ſine of HD: conſequently, the ſine of DCH is equal to half the ſine of BCH. Hence, the difference of the angles BCH, DCH being given, thoſe angles are given, and the arc CHI is given by poſition.

Moreover, in the triangle BCH, the baſe BH being biſected by the arc CD, the ſine of the angle CHD is to the ſine of the given angle CBD, as the ſine of the given angle HCD to the ſine of the given angle BCD; therefore, the angle CHB is given; in ſo much, that in the triangle CBH all the angles are given.

The ſum of the ſines of the angles BCH, DCH is to the difference of their ſines, as the tangent of half the ſum of thoſe angles to the tangent of half their difference; therefore, the tangent of half the ſum of BCH, DCH is three times the tangent of half BCD.

In (Fig. 6.) the iſoſceles triangle ABC, let the angle BAC be equal to the ſpherical angle BCD, and let AE be perpendicular to BC; alſo, CF being taken equal to CB, join AF: then EF is equal to three times EB; and as EF to EB, ſo is the tangent of the angle EAF to the tangent of EAB; but EAB is equal to half the ſpherical angle BCD: therefore, the angle EAF is equal to half the ſum of the ſpherical angles BCD, BCH; and conſequently, the angle CAF equal to the ſpherical angle DCH. Here, AF is to CF as the ſine of the angle ACF

2 to

to the ſine of C A F; and C B is to A B as the ſine of the angle B A C to the ſine of A C B: therefore, C F being equal to C B, and the ſine of A C F to the ſine of A C B, by equality, A F is to A B as the ſine of the angle B A C to the ſine of C A F, that is, as the ſine of the ſpherical angle B C D to the ſine of the ſpherical angle D C H.

Let (Fig. 7.) the triangle A G B have the angle A B G equal to the ſpherical angle C B D, and the ſide A G equal to A F. Then, A G is to A B as the ſine of the ſpherical angle B C D to the ſine of the ſpherical angle D C H, that is, as the ſine of the ſpherical angle C B H to the ſine of the ſpherical angle C H B: but A G is to A B alſo as the ſine of the angle A B G to the ſine of A G B; therefore, the angle A B G being equal to the ſpherical angle C B H, the angle A G B is equal to the ſpherical angle C H B: and moreover, when the angle A B G is greater than A B F, that is, when the ſpherical angle C B H is greater than the complement of half B C D, the three angles A B G, A G B and B A C together exceed two right.

Hence, (Fig. 8.) towards the equinoctial point C, where the angle C B D is obtuſe, a ſituation of the horizon, as B D, may always be found, wherein C D more exceeds C B than in any other ſituation: and when the acute angle D B A is greater than the complement of half B C D, another ſituation of the horizon, as K L M, may be found, toward the other equinoctial point A, wherein the arc of the ecliptic C K will be leſs than the arc of the equator, and their difference be greater than in any other ſituation. But, if the angle D B A be not greater than the com-

 plement

plement of half BCD, the arc of the ecliptic, between C and the horizon, will never be leſs than the arc of the equator, between the ſame point C and the horizon.

In the two ſituations of the horizon, the angles CHB and KMA are equal.

SCHOLIUM I.

To find the point in the ecliptic, where the arc of the ecliptic moſt exceeds the right aſcenſion, is a known problem: that point is, where the coſine of the declination is a mean proportional between the radius and the coſine of the greateſt declination.

In the preceding figure, ſuppoſing the angle CBD to be right, then, becauſe when CD moſt exceeds CB, the coſine of BD is to the radius as the ſine of CB to the ſine of CD, and, in the triangle CBD, the ſine of CB is to the ſine of CD as the ſine of the angle CDB to the radius, alſo the ſine of CDB is to the radius as the coſine of BCD to the coſine of BD; therefore, the coſine of BD is to the radius as the coſine of the angle BCD to the coſine of the ſame BD, and the coſine of BD is a mean proportional between the radius and the coſine of BCD.

SCHOLIUM 2.

In any given declination of the Sun, to find when the azimuth moſt exceeds the angle which meaſures the time from noon, is a problem analogous to the preceding.

Dr.

PROBLEM III.

*The tropic found, by Dr. Halley's method *, without any consideration of the parabola.*

The obſervations are ſuppoſed to give the proportions between the differences of the ſines of three declinations of the Sun near the tropic ; but the ſine of the Sun's place is in a given proportion to the ſine of the declination ; therefore, the ſame obſervations give equally the proportion between the differences of the ſines of the Sun's place, in each obſervation.

Now (Fig. 9.), let ACE be the ecliptic, AE its diameter between ♈ and ♎, and its center F ; let B, C, D be three places of the Sun ; BG, CI, DH the ſines of thoſe places reſpectively. Draw CK, BL parallel to AE, which may meet HD, in N and M. Then, by the obſervations, the ratio of DM to DN is given. Therefore, if BD be drawn to meet KL in O, the ratio of BD to OD is given ; and the ratio of BD to DC is alſo given, they being the chords of the given angles BFD, CFD: hence the ratio of CD to DO, in the triangle CDO, is given ; and conſequently, the angle COD will be given : which angle is the diſtance of the tropic from the middle point of the ecliptic between B and D : for, FPR being perpendicular to OC, and FQS perpendicular to DB, the angle QFP is equal to QOP, the points O, P, Q, F, being in a circle.

* Vide Philoſophical Tranſactions, N° 215.

THE

The Calculation.

$$\left.\begin{array}{l} DN : DM \\ s.\ \tfrac{1}{2}BFD : s.\ \tfrac{1}{2}CFD \end{array}\right\} :: \text{rad.} : t.\ \angle\ \chi$$

$$\text{rad.} : t.\ \angle\ \overline{\chi \backsim 45^\circ} :: t.\ \tfrac{1}{4}BFC : t.\ \frac{COD \backsim DCO}{2}$$

If $\chi > 45^\circ$, $\angle\ COD > DCO$

And

if $\chi < 45^\circ$, $\angle\ COD < DCO$.

If the intervals between the obſervations are ſo ſmall, that the ſines differ not much from the arches, the arches BC, CD may be counted in time, and the calculation may be abbreviated thus:

$DM : DN ::$ arc. $BD : Z$ (for DO)

$DC + Z : 2DC :: \tfrac{1}{2}BC : SR.$

Or,

$DM \times DC + DN \times BD : DM \times DC :: \tfrac{1}{2}BC : SR.$

XXXI. *On*

Received May 18, 1772.

XXXI. *On the Digeſtion of the Stomach after Death, by* John Hunter, *F. R. S. and Surgeon to* St. George's *Hoſpital.*

Read June 18, 1772.

AN accurate knowledge of the appearances in animal bodies that die of a violent death, that is, in perfect health, or in a ſound ſtate, ought to be conſidered as a neceſſary foundation for judging of the ſtate of the body in thoſe that are diſeaſed.

But as an animal body undergoes changes after death, or when dead, it has never been ſufficiently conſidered what thoſe changes are; and till this be done, it is impoſſible we ſhould judge accurately of the appearances in dead bodies. The diſeaſes which the living body undergoes (mortification excepted) are always connected with the living principle, and are not in the leaſt ſimilar to what may be called diſeaſes or changes in the dead body: without this knowledge, our judgment of the appearances in dead bodies muſt often be very imperfect, or very erroneous; we may ſee appearances which are natural, and may ſuppoſe them to have ariſen from diſeaſe; we may ſee diſeaſed parts, and ſuppoſe them in a natural ſtate; and we may ſuppoſe a circumſtance to have exiſted before

fore death, which was really a conſequence of it; or we may imagine it to be a natural change after death, when it was truly a diſeaſe of the living body. It is eaſy to ſee therefore, how a man in this ſtate of ignorance muſt blunder, when he comes to connect the appearances in a dead body with the ſymptoms that were obſerved in life; and indeed all the uſefulneſs of opening dead bodies depends upon the judgement and ſagacity with which this ſort of compariſon is made.

There is a caſe of a mixed nature, which cannot be reckoned a proceſs of the living body, nor of the dead; it participates of both, inaſmuch as its cauſe ariſes from the living, yet cannot take effect till after death.

This ſhall be the object of the preſent paper; and, to render the ſubject more intelligible, it will be neceſſary to give ſome general ideas concerning the cauſe and effects.

An animal ſubſtance, when joined with the living principle, cannot undergo any change in its properties but as an animal; this principle always acting and preſerving the ſubſtance, which it inhabits, from diſſolution, and from being changed according to the natural changes, which other ſubſtances, applied to it, undergo.

There are a great many powers in nature, which the living principle does not enable the animal matter, with which it is combined, to reſiſt, viz. the mechanical and moſt of the ſtronger chemical ſolvents. It renders it however capable of reſiſting the powers of fermentation, digeſtion, and perhaps ſeveral others, which are well known to

act

act on this ſame matter, when deprived of the living principle, and entirely to decompoſe it. The number of powers, which thus act differently on the living and dead animal ſubſtance, is not aſcertained: we ſhall take notice of two, which can only affect this ſubſtance when deprived of the living principle; which are, putrefaction and digeſtion. Putrefaction is an effect which ariſes ſpontaneouſly; digeſtion is an effect of another principle acting upon it, and ſhall here be conſidered a little more particularly.

Animals, or parts of animals, poſſeſſed of the living principle, when taken into the ſtomach, are not the leaſt affected by the powers of that viſcus, ſo long as the animal principle remains; thence it is that we find animals of various kinds living in the ſtomach, or even hatched and bred there: but the moment that any of thoſe loſe the living principle, they become ſubject to the digeſtive powers of the ſtomach. If it were poſſible for a man's hand, for example, to be introduced into the ſtomach of a living animal, and kept there for ſome conſiderable time, it would be found, that the diſſolvent powers of the ſtomach could have no effect upon it; but if the ſame hand were ſeparated from the body, and introduced into the ſame ſtomach, we ſhould then find that the ſtomach would immediately act upon it.

Indeed, if this were not the caſe, we ſhould find that the ſtomach itſelf ought to have been made of indigeſtible materials; for, if the living principle was not capable of preſerving animal

 ſubſtances

ſubſtances from undergoing that proceſs, the ſtomach itſelf would be digeſted.

But we find on the contrary, that the ſtomach, which at one inſtant, that is, while poſſeſſed of the living principle, was capable of reſiſting the digeſtive powers which it contained, the next moment, *viz.* when deprived of the living principle, is itſelf capable of being digeſted, either by the digeſtive powers of other ſtomachs, or by the remains of that power which it had of digeſting other things.

From theſe obſervations, we are led to account for an appearance which we find often in the ſtomachs of dead bodies; and at the ſame time they throw a conſiderable light upon the nature of digeſtion. The appearance which has been hinted at, is a diſſolution of the ſtomach at its great extremity; in conſequence of which, there is frequently a conſiderable aperture made in that *viſcus.* The edges of this opening appear to be half diſſolved, very much like that kind of diſſolution which fleſhy parts undergo when half digeſted in a living ſtomach, or when diſſolved by a cauſtic alkali, *viz.* pulpy, tender, and ragged.

In theſe caſes the contents of the ſtomach are generally found looſe in the cavity of the *abdomen*, about the ſpleen and diaphragm. In many ſubjects this digeſtive power extends much further than through the ſtomach. I have often found, that after it had diſſolved the ſtomach at the uſual place, the contents of the ſtomach had come into contact with the ſpleen and diaphragm, had

had partly diſſolved the adjacent ſide of the ſpleen, and had diſſolved the diaphragm quite through; ſo that the contents of the ſtomach were found in the cavity of the *thorax*, and had even affected the lungs in a ſmall degree.

There are very few dead bodies, in which the ſtomach is not, at its great end, in ſome degree digeſted; and one who is acquainted with diſſections, can eaſily trace the gradations from the ſmalleſt to the greateſt.

To be ſenſible of this effect, nothing more is neceſſary, than to compare the inner ſurface of the great end of the ſtomach, with any other part of the inner ſurface; what is ſound, will appear ſoft, ſpongy, and granulated, and without diſtinct blood veſſels, opaque and thick; while the other will appear ſmooth, thin, and more tranſparent; and the veſſels will be ſeen ramifying in its ſubſtance, and upon ſqueezing the blood which they contain from the larger branches to the ſmaller, it will be found to paſs out at the digeſted ends of the veſſels, and appear like drops on the inner ſurface.

Theſe appearances I had often ſeen, and I do ſuppoſe that they had been ſeen by others; but I was at a loſs to account for them; at firſt, I ſuppoſed them to have been produced during life, and was therefore diſpoſed to look upon them as the cauſe of death; but I never found that they had any connection with the ſymptoms: and I was ſtill more at a loſs to account for theſe appearances when I found that they were moſt frequent in thoſe who died of violent deaths, which made me ſuſpect that the true cauſe was not even imagined*.

At this time I was making many experiments upon digeſtion, on different animals, all of which were killed, at different times, after being fed with different kinds of food; ſome of them were not opened immediately after death, and in ſome of them I found the appearances above deſcribed in the ſtomach. For, purſuing the enquiry about digeſtion, I got the ſtomachs of a vaſt variety of fiſh, which all die of violent deaths, and all may be ſaid to die in perfect health, and with their ſtomach commonly full; in theſe animals we ſee the progreſs of digeſtion moſt diſtinctly; for as they ſwallow their food whole, that is, without maſtication, and ſwallow fiſh that are much larger than

* The firſt time that I had occaſion to obſerve this appearance in ſuch as died of violence and ſuddenly, and in whom therefore I could not eaſily ſuppoſe it to be the effect of diſeaſe in the living body, was in a man who had his ſkull fractured and was killed outright by one blow of a poker. Juſt before this accident, he had been in perfect health, and had taken a hearty ſupper of cold meat, cheeſe, bread, and ale. Upon opening the *abdomen*, I found that the ſtomach, though it ſtill contained a good deal, was diſſolved at its great end, and a conſiderable part of theſe its contents lay looſe in the general cavity of the belly. This appearance puzzled me very much. The ſecond time was at St. George's Hoſpital, in a man who died a few hours after receiving a blow on his head, which fractured his ſkull likewiſe. From thoſe two caſes, among other conjectures about ſo ſtrange an appearance, I began to ſuſpect that it might be peculiar to caſes of fractured ſkulls; and therefore, whenever I had an opportunity, I examined the ſtomach in every perſon who died of that accident: but I found many of them which had not this appearance. Afterwards I met with it in a ſoldier who had been hanged.

the

the digeſting part of the ſtomach can contain (the ſhape of the fiſh ſwallowed being very favourable for this enquiry,) we find in many inſtances that the part of the ſwallowed fiſh which is lodged in the digeſting part of the ſtomach is more or leſs diſſolved, while that part which remains in the *œſophagus* is perfectly ſound.

And in many of theſe I found, that this digeſting part of the ſtomach was itſelf reduced to the ſame diſſolved ſtate as the digeſted part of the food.

Being employed upon this ſubject, and therefore enabled to account more readily for appearances which had any connection with it, and obſerving that the half-diſſolved parts of the ſtomach, &c. were ſimilar to the half-digeſted food, it immediately ſtruck me that it was from the proceſs of digeſtion going on after death, that the ſtomach, being dead, was no longer capable of reſiſting the powers of that menſtruum, which itſelf had formed for the digeſtion of its contents; with this idea, I ſet about making experiments to produce theſe appearances at pleaſure, which would have taught us how long the animal ought to live after feeding, and how long it ſhould remain after death before it is opened; and above all, to find out the method of producing the greateſt digeſtive power in the living ſtomach: but this purſuit led me into an unbounded field.

Theſe appearances throw conſiderable light on the principles of digeſtion; they ſhew that it is not mechanical power, nor contractions of the ſtomach, nor heat, but ſomething ſecreted in the coats of the ſtomach,

ſtomach, which is thrown into its cavity, and there animaliſes the food *, or aſſimilates it to the nature of the blood. The power of this juice is confined or limited to certain ſubſtances, eſpecially of the vegetable and animal kingdoms ; and although this menſtruum is capable of acting independently of the ſtomach, yet it is obliged to that *viſcus* for its continuance.

* In all the animals, whether carnivorous or not, upon which I made obſervations or experiments to diſcover whether or not there was an acid in the ſtomach, (and I tried this in a great variety,) I conſtantly found that there was an acid, but not a ſtrong one, in the juices contained in that *viſcus* in a natural ſtate.

XXXII. *Experiments and Obſervations on the Waters of* Buxton *and* Matlock, *in* Derbyſhire, *by* Thomas Percival, *of* Mancheſter, *M. D. and F. R. S.*

Read June 25, 1772.

THE water of ſaint Ann's-well is found, by analyſis, to contain calcareous earth, foſſil- alkali, and ſea ſalts; but in very ſmall proportions: for a gallon of the water, when evaporated, yields only twenty three, or twenty four grains of ſediment. It ſtrikes a light green colour with ſyrup of violets, ſuffers no change from an infuſion of galls, from the fixed vegetable alkali, or from the mineral acids; becomes milky with the volatile alkali, and with Saccharum Saturni; and lets fall a precipitate on the addition of a few drops of a ſolution of ſilver, in the nitrous acid. The ſpecific gravity of this water is preciſely equal to that of rain water, when their temperatures are the ſame; but it weighs four grains in a pint lighter, when firſt taken from the ſpring. The heat of the bath is about 82 degrees of Fahrenheit's thermometer; that of Saint Ann's well, as it is a ſmaller body of water, and expoſed to the open air, is ſomewhat leſs. The water is tranſparent, ſparkling, and highly grateful to the palate *.

* I am indebted to the information of the worthy phyſician who attends at Buxton, for ſome of theſe facts.

In

In October 1769, I paſſed a few days at Buxton; and during my ſtay there amuſed myſelf with the following experiments on the effects of the water of Saint Ann's well, on my pulſe.

EXPERIMENT I.

October 12, eight o'clock in the morning. The day cold and moiſt, my pulſe beat 84 ſtrokes in a minute; I drank at the well, the third of a pint of water, and, uſing every neceſſary precaution, examined my pulſe at certain intervals of time; in five minutes, pulſe 80, in ten minutes pulſe 80, fuller and harder; in twenty minutes pulſe 85; in half an hour pulſe 90.

EXPERIMENT II.

Eleven o'clock in the forenoon, two hours after breakfaſt, the air warm and ſerene, pulſe 90; I repeated the draught of water. In ſeven minutes pulſe 109; in fifteen minutes pulſe 103; in thirty minutes pulſe 100, head-ach; in an hour and a half pulſe 95, head-ach abated.

EXPERIMENT III.

October 13, eight in the morning; the day cold, pulſe 92; I drank the quantity of water above-mentioned; in five minutes pulſe 86; in fifteen minutes pulſe 86, full and hard; in twenty minutes pulſe 100; in half an hour pulſe 92.

From the firſt and third experiments, it appears that the coldneſs of the morning counteracted for a time, the effects of the Buxton water; and reduced the

the vibrations of my pulſe from 84 to 80, and from 92 to 86. But the ſtimulus of the water ſoon became ſuperior to the ſedative powers of the cold to which I was expoſed; for within the ſpace of half an hour my pulſe roſe to 90 in the firſt, and to 100 ſtrokes in the ſecond trial. At eleven o'clock before noon, when the air was warm and ſerene, the water in a much ſhorter time excited its force, increaſing the velocity of my pulſe from 90, to 109 vibrations in a minute. Theſe experiments evince the heating quality of Buxton water, and ſuggeſt to us the precautions to be obſerved in the uſe of it. Small quantities ſhould only be drunk at once, and frequently repeated; the belly ſhould be kept ſoluble with lenitive Electuary, or any other mild purgative and at the beginning of the courſe, the patient may be directed to ſuffer the water to remain a few ſeconds in the glaſs, before he ſwallows it. For this celebrated ſpring abounds with a mineral ſpirit, or mephitic air, in which its ſtimulus, and indeed its efficacy reſides, and which is quickly diſſipated by expoſure to the air.

The honourable and ingenious Mr. Cavendiſh has ſhewn by his Experiments on Rathbone Place water, Ph. Tranſactions, vol. LVII, that calcareous earths may be rendered ſoluble in water, by furniſhing them with more than their natural property of fixed air. And it has lately been diſcovered that iron alſo may be ſuſpended by this principle, in the ſame menſtruum *. It appeared therefore highly probable to me, that a chalybeate impregnation might with great facility

* Vid. Mr. Lane's experiments, Ph. Tranſactions, Vol. LIX.

be communicated to the Buxton water, when fresh drawn from the spring; a quality, which in many cases would add greatly to its medicinal efficacy. I suggested the trial to Mr. Buxton, a very worthy and sensible apothecary near the wells, who has lately at my request made the following experiment.

Experiment IV.

A quart bottle containing two drachms of iron filings, was filled by immersion, with the water of Saint Anne's well, corked and agitated briskly under the surface of the water: it was then suffered to remain in the well till the filings had subsided, when the water was carefully decanted into a half pint glass; to this were added three drops of the tinctureof galls, which immediately occasioned a deep purple colour, and transparency was presently restored by a few drops of the acid of vitriol; evident proofs that a solution of the iron was effected in a few minutes. The water also without the tincture of galls had a chalybeate taste, and left an agreeable astringency on the palate.

By this experiment, it appears that a warm chalybeate abounding with a mineral spirit, and grateful to the taste, may with very little trouble be obtained. And this method of impregnating the Buxton water with iron, must increase its tonic powers, and in many cases improve its medicinal virtues. It is a common practice to join the use of a chalybeate spring in the neighbourhood of St. Anne's well, with that of the Buxton water: but, the superiority of the artificial mineral water must be apparent, if we consider its agreeable warmth, volatility, levity, and gratefulness to the palate.

Buxton.

Buxton bath is very frequently employed as a temperate cold bath. For as the heat of the water is about ſixteen or eighteen degrees below that of the human body, a gentle ſhock is produced on the firſt immerſion, the heart and arteries are made to contract more powerfully, and the whole ſyſtem is braced and invigorated. But this ſalutary operation muſt be greatly diminiſhed, often indeed more than counter-balanced, by the relaxing vapours which copiouſly exhale from the bath, to which the patients are expoſed during the time of dreſſing and undreſſing. A ſeparate room is indeed provided for the ladies; but the gentlemen have no other accommodations than what the vault affords in which the bath is contained, and are therefore liable to all the inconveniences ariſing from warmth and moiſture. June 12, 1772, the mercury ſtood in the ſhade at 65, but in this vault quickly aroſe to 78 degrees.

Experiments on MATLOCK WATER.

Experiment I.

A thermometer made by Dollond, and graduated according to Fahrenheit's ſcale, was expoſed for a ſufficient length of time, to the ſteam of the water, as it guſhes from the rock, and alſo immerſed in the baſon that receives it. The mercury roſe to 66 degrees.

Experiment II.

Six drops of Sp. Sal. Ammon. vol. were poured into a glaſs of the ſpring water, which contained about the ſixth of a pint; a very ſlight cloudineſs immediately enſued, but no precipitation was afterwards obſervable.

Experiment III.

Six drops of a ſolution of ſalt of tartar occaſioned a cloudineſs, juſt perceptible, in the ſame quantity of water; no precipitation enſued.

Experiment IV.

Six drops of a ſolution of ſaccharum ſaturni immediately produced a milkineſs in the water, but no ſenſible precipitation.

Experiment V.

Six drops of a ſolution of ſilver in the nitrous acid inſtantly occaſioned a milkineſs in the water; and after ſtanding an hour, a grey powder was obſervable at the bottom of the glaſs.

Experiment VI.

Ten drops of the infuſion of galls neither produced any change of colour in the water at the time they were added, nor was the ſlighteſt purple hue perceptible two hours afterwards.

Experiment VII.

A piece of paper beſmeared with ſyrup of violets was dipped into a glaſs full of water; no change of colour enſued.

EXPERIMENT VIII.

Another piece of paper, moiſtened in the ſame manner with the ſyrup, was placed over a glaſs of water, as ſoon as it was taken from the ſpring. The paper ſuffered no change of colour, although it remained an hour upon the glaſs.

EXPRRIMENT IX.

My pulſe beat 84 ſtrokes in a minute, at the time when I drank a half pint glaſs of the Matlock water; in 20 minutes my pulſe roſe to 86; in half an hour after they ſunk to 82, and continued to vibrate the ſame number of times for an hour, which was as long as I thought it was neceſſary to examine them.

EXPERIMENT X.

The mercury in the thermometer, when immerſed in each of the baths, ſtood at 68: in the river Derwent, which flows through the valley of Matlock, at 52. Theſe experiments were made in the month of June 1772, and the weather was warm.

EXPERIMENT XI.

A four ounce phial, after being accurately counterpoiſed in a very nice balance, was filled to the brim with diſtilled water, which weighed three ounces, four drachms, forty five grains and a half. The ſame phial, exactly balanced as before, was then filled to the brim with Matlock water, of the ſame temperature

perature with the diſtilled water, which weighed three ounces, four drachms, and forty-ſix grains.

Matlock water is grateful to the palate, and of an agreeable temperature, but exhibits no marks of any mineral ſpirit, either by its taſte, ſparkling appearance in the glaſs, or by the chemical teſt employed in experiment 8. The ſecond and third experiments ſhew that it is very ſlightly impregnated with Selenites or other earthly ſalts; and of this its comparative levity affords alſo a further proof: for it weighs twenty-ſix grains in a pint lighter than the Mancheſter pump water*, and only four grains heavier than diſtilled water. The precipitation of a grey powder, by the adding of a ſolution of ſilver in aqua fortis to the water, renders it probable that a ſmall portion of ſea ſalt is contained in it. For the powder is found to conſiſt of the particles of ſilver, combined with the muriatic acid, which is ſeparated from the foſſil alkali by the ſuperior affinity the nitrous acid bears to it; and thus a double elective attraction takes place in this experiment.

This water is ſaid to contain iron, but the aſſertion is at leaſt rendered doubtful by the 6th experiment, which was made with the utmoſt accuracy; and I am inclined to think, that it is entirely without foundation. The ſpring is juſtly celebrated for its efficacy in hæmoptoes; and hence it may have been too haſtily concluded that it poſſeſſes ſome ſlight degree of ſtypticity, by means of a chalybeate impregnation.

* Vid. the author's treatiſe on the pump water of Mancheſter. Eſſays medical and experimental, p. 207. 2d edit.

The

The 9th experiment, which my ſhort ſtay at Matlock would not allow me leiſure to repeat, affords a preſumption that the water is not poſſeſſed of any ſtimulating powers; for the ſmall increaſe of quickneſs in my pulſe, on drinking half a pint of it, may be aſcribed more to the quantity received into the ſtomach, than to the heating quality of the water,

The Briſtol and Matlock waters appear to reſemble each other, both in their chemical and medicinal qualities. I have examined and compared them together by the teſt mentioned above, and ſo far as ſuch trials may be be deemed concluſive, there ſeems to be no other than the following ſlight difference between them,

Briſtol water becomes a little more milky on the addition of a ſolution of fixed alkali, and of Saccharum Saturni than that of Matlock; the former alſo weighs near a grain in a pint heavier than the latter. Is it not to be lamented therefore, that ſo little attention is paid to Matlock, even by the phyſicians who reſide in the neighbourhood of it? In hectic caſes, hæmoptoes, the diabetes, and other diſorders, in which the circulation of the blood is rapid and irregular, I ſhould apprehend that Matlock water, on ſome accounts, claims the preference to that of Briſtol; for it is leſs diſpoſed to quicken the pulſe, and may therefore be drunk in larger quantities. But it muſt be acknowledged that the climate of Briſtol is ſuperior to that of Matlock, a circumſtance of the higheſt importance to conſumptive patients. Situated in a deep though delightful valley, and ſurrounded by very high mountains, the ſun diſappears at Matlock earlier in the evenings, the fogs are longer in diſperſing, and it may be preſumed that rain falls here more frequently and copiouſly than in other places. For at Catſworth, which is encompaſſed alſo with hills, and is about ten miles diſtant, in 1764, 1765, 1767, and 1768, about 33 inches of rain fell at a medium each year.

The following table exhibits a comparative view of the different temperatures of Bath, Buxton, Briſtol, and Matlock waters, meaſured by Fahrenheit's thermometer.

* BATH.	
King's Bath Pump	112
Hot Bath Pump	114½
Croſs Bath Pump	110
* BRISTOL.	
Hot Well Pump	76
BUXTON.	
Bath	82
St. Ann's Well	81 ×
MATLOCK.	
Baths	68
Spring	66

* Vid. Mr. Canton's experiments. Ph. Tranſ. Vol. LVII. p. 203.

XXXIII. *Som*

XXXIII. *Some Account of a Body lately found in uncommon Preservation, under the Ruins of the Abbey, at* St. Edmund's-Bury, Suffolk; *with some Reflections upon the Subject: By* Charles Collignon, *M. D. F. R. S. and Professor of Anatomy at* Cambridge.

Read June 25, 1772.

IN the month of February last, some workmen, digging among the ruins of the above abbey, discovered a leaden coffin, supposed, from some circumstances, to contain the remains of Thomas Beaufort, Duke of Exeter, uncle to king Henry the Fifth. As it certainly was buried before the dissolution of the abbey, it must have been there between two and three hundred years. It was found near the wall, on the left-hand side of the choir of the chapel of the blessed Virgin; not inclosed in a vault, but covered over with the common earth. Upon examining the appearance of the body, the following circumstances were remarkable, as communicated to me, by an ingenious surgeon, on the spot, Mr. Thomas Cullum.

"The body was inclosed in a leaden coffin, surrounding it very close, so that you might easily distinguiſh the head and feet. The corpſe was wrapped round with two or three large layers of cere-cloth, ſo exactly applied to the parts, that the piece, which covered the face, retained the exact impreſſion of the eyes and noſe. The dura mater was entire. The brain was of a dark aſh colour, with ſome remaining appearance of the medullary part. The coats of the eye were ſtill whole, and had not totally loſt their gliſtening appearance. There was about half a pint of a bloody-black water in the thorax; and a maſs that ſeemed to be part of the lungs. The pericardium and diaphragm were quite entire. The abdominal viſcera had been taken out very clean, and the integuments and muſcles ſtuck very cloſe to the vertebræ of the back. This cavity looked freſher than that of the thorax. I cut into the pſoas magnus, where there were evident marks of red muſcular fibres. The other muſcles had loſt all their red colour, and were become of a dark brown. The tendons were ſtill ſtrong, and retained their natural appearance. The hands, which are preſerved in ſpirits, retain the nails. There were ſome very ſmall holes in the coffin, out of which had run ſome bloody water, of an offenſive ſmell. All the principal blood-veſſels muſt have been cut through, in taking out the abdominal viſcera: and if no ligature was made upon the veſſels, their contents would eſcape, particularly as aſſiſted by the preſſure of the cere-cloth, which is of conſiderable weight, and, doubtleſs, put on hot. This fluid running out of the coffin, upon its being moved, might occaſion the ſuſpicion of the body being put in pickle."

Thus

Thus far Mr. Cullum's account, by which it appears, that the viscera of the abdomen had been taken out, so that the greatest part of the blood, he observes, did probably flow out, during that operation, from the mouths of the divided vessels, and whose diameter is considerable. This would greatly reduce the quantity of the fluids. The holes in the coffin, if purposely made, would seem designed to let out extravasated or transuding fluids; but are irreconcileable with the notion of the body being in pickle. If the holes were accidental, the notion of a pickle may still be allowed. Might not the cere-cloth, impregnated, perhaps, with gums or resins, and, from its taking so exact an impression, most probably laid on hot preclude the external air; and, if done immediately after the party's death, obviate the deposition of eggs, or incapacitate them from ever hatching? The lead grasping close, would co-operate with the cere-cloth in the exclusion of air and insects.

We have undoubted accounts of bodies found very little changed, after long interment, where there was no appearance of any art having been used. And there is no doubt some constitutions are more prone to putrefaction after death than others; these circumstances may be dependant on the age, sex, and last disease; to which predisposing causes, thus attending persons to the grave, are to be added the soil and situation in which they are deposited. Could we be masters of all these particulars, in the few dead bodies hitherto discovered greatly free from the usual putrefaction, it would lead, perhaps, to the probable cauſe of the phænomenon, and point out a proper method of imitation. And till that is done, it is difficult to know how much merit is to be aſſigned to the art or myſtery of embalming, and how much to the power of natural cauſes.

XXXIV. *A Letter from* Richard Pulteney, *M. D. F. R. S. to* William Watson, *M. D. F. R. S. concerning the medicinal Effects of a poisonous Plant exhibited instead of the Water Parsnep.*

Dear Sir,

Read July 9, 1772.

SOME circumstances having lately come to my knowledge, relating to the effects of a poisonous plant, I thought them rather too remarkable not to merit further notice; and, I address them to you with the more propriety, as you have already laid before the publick some observations * concerning the deleterious qualities of the plant in question, which holds a distinguished place among the poisonous ones that are indigenous in Britain.

Mr. H——n, an attorney of this place, now upwards of forty, at the age of fifteen, began to be affected (after taking cold upon violent exercise, as he thinks) with what is usually called a scorbutick disorder; which shewed itself more particularly on the outsides of his arms, about the elbows, and on

* See Philosophical Transactions, Vol. XLIV. p. 227. and Vol. L. p. 856.

the

the outſides of his legs, from the knees to the ancles, as well as in blotches upon other parts of his body. It had the appearance of a dry branny ſcab or ſcurf, which every night fell off, more or leſs, in ſcales, as is uſual in leprous caſes. At times it puſhed out more than uſual, and thickened the integuments of the limbs conſiderably, after which the ſeparation of ſcales would become very abundant.

For ſeveral years paſt he had been trying a variety of things commonly recommended in ſuch caſes, particularly the quack medicine known by the name of Maredant's Drops, which he continued for near a twelvemonth, without finding the leaſt ſenſible relief: alſo an electuary of Flos ſulphuris and Cremor tartari, which he had perſevered in for near three years, without finding any other alteration, than that of its preventing coſtiveneſs, to which he was habitually ſubject.

In the winter 1770, this diſorder increaſed upon him very rapidly, without being able to aſſign any reaſon, from any accident that had happened to him, or from any irregularity of his own in point of regimen, in which he was always very exact. At this time, beſides the farther ſpreading of the eruption itſelf, the integuments of the legs thickened very much, and the limbs ſwelled to ſuch a degree, as to render him unable to walk. The quantity of branny ſcurf and ſcales thrown off, at this time, was very great; he ſays "handfuls might have been taken out of his bed every morning."

In this unhappy ſituation, even loathſome to himſelf, it was recommended to him to take the juice of water parſnep, in the quantity of one common table-

ſpoonful

ſpoonful every morning, faſting, mixed with two ſpoonfuls of white mountain wine.

Accordingly, about the middle of January 1771, he procured a half-pint phial of what was ſo called, by means of the perſon who had recommended it, and who had aſſured him that he had been greatly relieved, in a ſimilar diſorder, by it.

The firſt ſpoonful he took did not begin to give any great uneaſineſs for two hours, but after that time, his head began to be affected in a very extraordinary manner; a violent ſickneſs ſoon ſucceeded, and violent vomiting; and, after he was put to bed, there came on cold ſweats, and a very ſtrong and long-continued rigor, ſo that the people about him thought him dying for ſome time; but, in a few hours, all theſe ſymptoms wore off.

Such, however, had been the inveteracy of his diſorder, and ſo ſtrong his deſire to find relief, that he determined not to deſiſt; and, after having omitted his medicine for one day, he repeated it, in nearly the ſame doſe, and with ſimilar effects as to ſickneſs and vomiting, though the uncommon ſenſation in his head, and the ſucceeding rigor, were by no means ſo violent. He had reſolution enough to continue this doſe every other morning, for more than a fortnight, and then reduced it to three tea-ſpoonfulls which was juſt the half of the firſt doſe.

Before he had taken this juice one month, he was ſenſible of a very great change for the better; encouraged, therefore, by theſe appearances he perſevered in its uſe until the middle of April, by which time his ſkin, though not quite cleared, yet had ceaſed to throw off any more ſcurf, was become

come

come ſoft, clean, and well conditioned, and, as he has repeatedly aſſured me, he got then into a much better conditioned ſtate, then he had experienced for many years before.

From firſt to laſt, this juice never purged him; though he ſays, even in its reduced doſe, it never failed to occaſion a dizzineſs of the head, a nauſea, and ſickneſs, which were not infrequently ſucceeded by a vomiting, that always inſtantly relieved his head.

From the middle of April to the middle of June, he deſiſted from the uſe of the juice, but, in its ſtead, drank every morning for breakfaſt, the infuſion of the leaves of the ſame plant, which, he ſays, is like common bohea tea. The infuſion ſeldom occaſioned nauſea, or ſickneſs, but always brought on a ſmall degree of vertigo, and in a ſlight manner produced the effects of intoxication from liquor.

In June he went to Harrowgate, as he had deſigned in the ſummer before. Upon firſt drinking and bathing there, he thought himſelf worſe; and his eruptions, having gradually increaſed during the two months that he ſtaid in that place, he was convinced that thoſe waters were of no real ſervice to him. On his coming home, he returned to the uſe of the infuſion, and he aſſures me, that he again found, even by that weak preparation, a very ſpeedy alteration for the better. From that time, he continued it ever ſince, until his ſtock of the herb was exhauſted; his ſkin is now ſo very little affected, that he has but here and there, upon his arms and legs, a very ſmall appearance of his diſorder.

Upon queſtioning him relating to the ſenſible qualities of this medicine, he ſays again, that he

part-

particularly remembers that it never once purged him; not even the first dose, which had so nearly poisoned him. He does not think that it increased the sensible perspiration, but is convinced that it was diuretick; and adds, that he thinks it occasioned, besides the increased flow of urine, a copious sediment in it, and which he believes was always wanting before.

This is the plain, narrative of the fact. He has assured me that no medicine or regimen, among the great variety that he has tried, ever had any sensible effect upon his disorder before; and that nothing but the very early and sensible relief he experienced from this juice, could have induced him to persevere in its use, under such uneasy feelings, as it never failed to produce. Indeed, he makes nothing of the lighter effects of the infusion, from which, however, he thinks, he has likewise reaped no small benefit.

This case, the nature and inveteracy of his disorder, being well known among his neighbours, was much talked of, and raised the curiosity of many people. When I first heard of it, and was informed of the smallness of the dose, and its virulent operation, I could scarce doubt that the juice of some other plant had been administered instead of that of the water parsnep, which we know to be a safe and harmless vegetable; medical writers having directed its juice to be drunk, even to the quantity of four ounces for a dose: and as I know, the *Oenanthe crocata*, hemlock dropwort, to be exceedingly plentiful in this country, so much, as to be more easily procured than the water parsnep itself; I thought it

 probable

probable that that plant had been uſed in its ſtead. Upon getting a ſpecimen, it appeared that this had been indeed the caſe; as alſo, upon farther enquiry, that it was the juice of the root only, and not of the leaves and ſtalks, that had been adminiſtered. I might here obſerve, that the expreſſion from the root is not to be depended upon after the plant is advanced towards its flowering ſtate, as the root then becomes light, ſpungy, and almoſt deſtitute of juice.

If you judge this caſe not improper to be laid before the Royal Society, you will do me the honour of preſenting it. Mr. H——n himſelf is ſo much convinced of the efficacy of the medicine, that he is deſirous of its being known to the world.

I do not enter into any reaſoning on this occurrence; I relate it only as a fact, and deſire it may have no more weight than every judicious phyſician knows is due to a ſingle inſtance. How far it may be proper to give this juice a farther trial, I will not take upon me to determine; but muſt, as an encouragement to any who may chuſe to venture upon it, inform them, that it has not on all perſons ſo much power in producing nauſea and ſickneſs, as in the caſe here before us. I am,

S I R,

with great eſteem,

Your obliged humble ſervant,

Blandford,
March 12, 1772.

R. Pulteney.

P. S.

7

P. S. Mr H——— is desirous that it should be known, that he " tried very fruitlesly, among other methods, the drinking of tar-water and sea-water, of each of which, he says, he did not drink less than an hogshead."

XXXV. April 21, 1772. *Experiments on two Dipping-Needles, which Dipping-Needles were made agreeable to a Plan of the Reverend Mr.* Mitchell, *F. R. S. Rector of* Thornhill *in* Yorkshire, *and executed for the Board of Longitude, by Mr.* Edward Nairne, *of* Cornhill, London.

Read July 9, 1772. THE magnetic needles were twelve inches long, and their axes (the ends of which were of gold allayed with copper) rested on friction-wheels of four inches diameter, each end on two friction-wheels, which wheels were balanced with great care. The ends of the axes of the friction-wheels were likewise of gold allayed with copper, and moved in small holes made in bell-metal; and opposite the ends of the axes of the needles, and the friction-wheels, were flat agates, finely polished. Each magnetic needle vibrated in a circle of bell-metal, divided into degrees and half-degrees, and a line passing through the middle of the needle to the ends pointed to the divisions. The minutes set down in the experiments were, by estimation, as the third of half a degree is counted ten minutes. The instruments were carefully placed, so that the needles vibrated exactly in the magnetic meridian.

E

40

30

20

20

This Scale 5/8 to one Inch.

1

12

J. Basire sc.

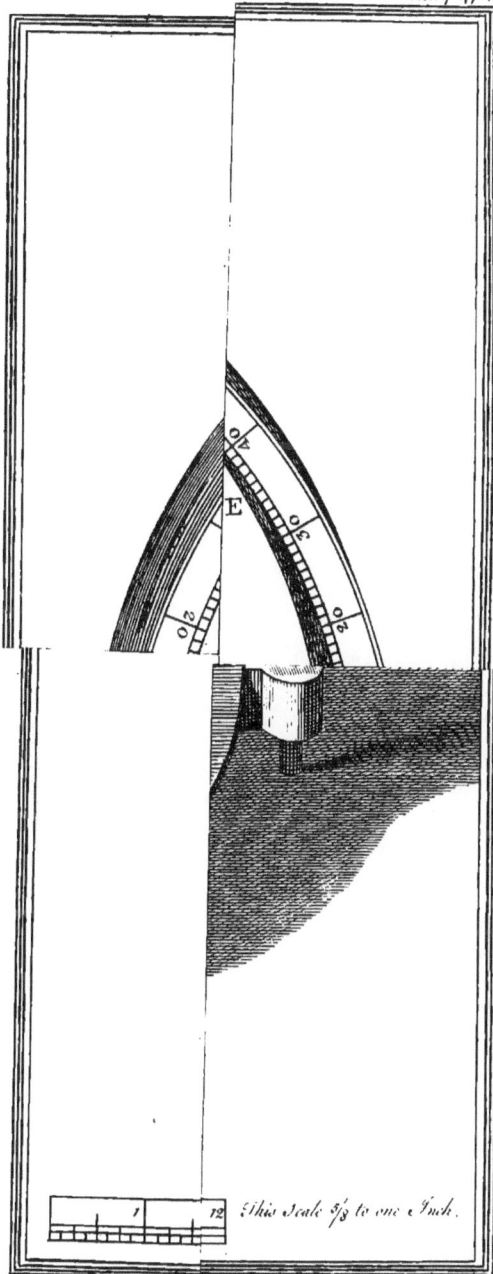

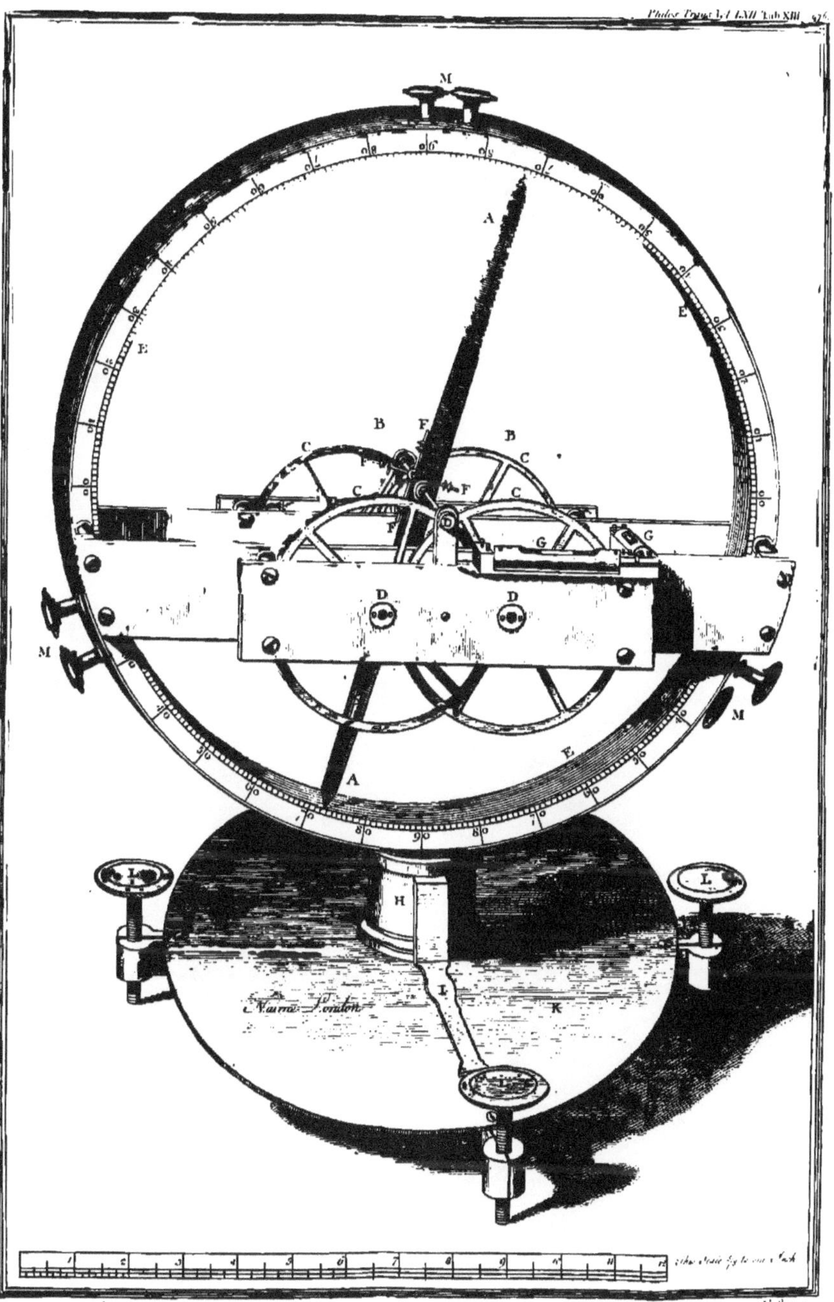
M
A
E
B
F
C
D
G
H
L
I
K
Nairne London

meridian. The two needles were nearly balanced before they were made magnetical; but, by a curious contrivance of the Reverend Mr. Mitchell of a croſs fixed on the axes of the needles (on the arms of which were cut very fine ſcrews, to receive ſmall buttons, that might be ſcrewed nearer or farther from the axis), the needles could be adjuſted both ways, to a great nicety, after they were made magnetical, by reverſing the poles, and changing the ſides of the needle.

Firſt ſet of experiments made by Edward Nairne, at his houſe, N° 20, Cornhill.

°	′
72	20
72	20
72	20
72	20
72	20
72	20.

Second ſet of experiments, with that ſide of the inſtrument to the Eaſt, which was to the Weſt in the firſt obſervation.

°	′	
72	10	
72	15	
72	45	Here the ends of the axis touched the agates.
72	45	
72	5	
72.		

Third

Third ſet of experiments, in which the poles of the needle were reverſed, but the ſame ſide of the inſtrument to the Eaſt, as in the ſecond ſet of experiments, and the needle rather more magnetical, being touched with a larger ſet of magnets.

°	′
72	30
72	30
72	30
72	30
72	30
72	30.

Fourth ſet of experiments, viz. the ſame ſide of the inſtrument to the Eaſt, as in the firſt ſet of experiments.

°	′	
72	10	
72	10	
72	15	Obſerved by Mr. Wales.
72	10	
72	10	
72	10.	

Fifth experiment, viz. the ſame end of the needle made North, as in the firſt ſet of experiments, and alſo the ſame ſide of the inſtrument to the Weſt, as in the firſt ſet of experiments.

°	′
72	20.

Experiments

Experiments made April 22, 1772, with the other Dipping-needle, the inſtrument being put in the ſame place, and with great care, in the magnetic meridian, the needle pointed as under.

°	′	
72	15	
72	10	The poles of the needle changed.
72	20	The ſide of the inſtrument to the Eaſt, which in the firſt obſervation was to the Weſt.

Leſt any thing magnetical ſhould have affected the needle in Mr. Nairne's houſe, he took this inſtrument, and placed it in the middle of a large room belonging to the London Aſſurance in Birchin-Lane, and then the needle pointed to

°	′	
72	10 or 15	
72	20	
72	30	The poles of the needle changed.
72	10	The ſide of the inſtrument to the Eaſt, which in the firſt obſervation was to the Weſt.

The dipping-needle brought back to Mr. Edward Nairne's, and put in the ſame place as before, ſtood at

°	′
72	10 +

The

In the foregoing experiments, the needle was raiſed to an horizontal poſition, and left to vibrate. It was between 8 or 9 minutes before the vibration ceaſed.

The needle brought to an horizontal poſition, and one grain and a half laid on the extremity of the South end, was not ſufficient to keep it in an horizontal poſition; but the North end pointed to 35° 30′. One grain and three quarters laid on the extremity of the South end of the needle, was more than ſufficient to keep it in an horizontal poſition, the South end then pointing 6° 45′ below o.

It having been judged proper to have a Drawing of the Dipping-Needle, the following Plate [TAB. XIII.] has been made, wherein

A A Repreſents the needle.
B B The ends of the axis reſting on the friction-wheels.
CCCC The four friction-wheels.
DDD Where flat agate caps are ſet in.
EEE The divided circle of bell-metal.
FFFF The ends of the croſs for adjuſting the needle.
GG Two levels, whereby the line of o degrees of the inſtrument is ſet horizontal.
H The perpendicular axis, whereby the inſtrument may be turned, that the divided face of the circle may front the Eaſt or Weſt.
I An index fixed to the perpendicular axis H, and which points to an oppoſite line on the horizontal plate K, when the inſtrument is turned half round.
LLLL Four adjuſting ſcrews to ſet the inſtrument horizontal. One of them is hid behind the circle.
MMMM Screws which hold on the glaſs covers, to keep the needle from being diſturbed by the wind.

INDEX

AN

INDEX

TO THE

Sixty-Second VOLUME

OF THE

Philosophical Transactions.

A.

B.

C.

D.

E.

F.

G.

George

H.

I.

K.

Land-rail,

L.

M.

N.

O.

P.

Q.

R.

S.

T.

V.

W.

Water,

5

The End of the SIXTY-SECOND VOLUME.

*** There are FOURTEEN Copper-Plates in this Volume, as Table IV. is double.

E R R A T A.

Vol. LXI.

Pag. 139. line 11. *from the bottom, read* upon, with regard to
141. l. 1. *notes, erase the comma after* Ex,
143. *notes*, l. penult. *r.* Archiepiſcopis. l. 15. *r.* Redleiam
144, l. 2, *r.* Dena. *Notes*, l. 14. *from the bottom, r.* Noewera, l. ult. *r.* Vincentii.
145. *notes*, l. 4. *r.* Crevecor.
147. l. 3. *the 4th letter in the Saxon word ſhould be* ꞇ.

Vol. LXII.

Pag.	line	*for*	*read*
xi.	line penult.	vingtimee	vingtieme
6.	6	Caniculus	Cuniculus
8.	1.	male	mule
ibid.	14.	is in other	is other
37.	7.	Juptiter	Jupiter
55.	21.	grows	it grows
75.	21.	diſtantis	diſtantia
77.	22.	(Tab. IV.)	(Tab.IV.&Tab.IV.*)
125.	note †, l. 4.	weter	water
146.	8,	them	it
303.	note *, l. 2.	Aëdologue	Aëdologie
314.	17.	cough	chough
388.	21.	Three-toid	Three-toed
426.	17.	veſti	veſtiti
429.	6.	mandibu	mandibula
457.	27.	property	proportion
462.	note, line laſt,	207	287.